# Corporate Sustainability Disclosure Matters

Timo Heroth

# Corporate Sustainability Disclosure Matters

## Evidence from the Field

Springer Gabler

Dr. Timo Heroth
Betriebswirtschaftslehre
Universität St. Gallen
Riehen, Switzerland

This book is a reprint of the dissertation accepted by the University of St. Gallen.

ISBN 978-3-658-47652-6 ISBN 978-3-658-47653-3 (eBook)
https://doi.org/10.1007/978-3-658-47653-3



This Springer Gabler imprint is published by the registered company Springer Fachmedien Wiesbaden GmbH, part of Springer Nature.
The registered company address is: Abraham-Lincoln-Str. 46, 65189 Wiesbaden, Germany

# Foreword

The question of the legitimacy of companies is one of the fundamental questions of business administration, although it should not be overlooked that the concept of legitimacy and legitimacy theory can be understood, classified and interpreted in very different ways. Essentially, however, there is agreement that—as Meyer/Rowan already noted in 1977—the legitimacy of an organization is decisive for its ability to survive or continue as a going concern. The possible answers to the question of legitimacy are therefore correspondingly diverse: on the one hand, because they depend to a large extent on the underlying theories that are considered relevant. On the other hand, and related to this, because the answers change over time. For example, the purpose of companies may be seen as increasing shareholder value or, in a broader context, as increasing stakeholder value or—even more broadly—as creating value for society as a whole, so-called public value. Depending on which theory or approach is followed, there are sometimes completely different consequences, for example for the regulatory targets to be pursued, the specific structure of corporate reporting or the financial and sustainable performance of companies.

This dissertation by Mr. Heroth—submitted to the University of St. Gallen under the title "The Value Relevance of Corporate Stakeholder Orientation and Public Value Topic Salience—Measuring the Impact of Discursive Shifts in Society with the help of Corporate Sustainability Disclosure"—is dedicated to this major, fundamental topic outlined above. In essence, it is about the exciting analysis of the extent to which a stakeholder orientation of companies, with special consideration of social issues, can have an influence on financial and sustainable performance. Such an analysis is particularly relevant for a variety of reasons:

Firstly, the importance and topicality of work conceptually based on stakeholder orientation and the public value approach can hardly be overestimated at present, as regulatory developments—in particular those in the European Union (keyword: "Green Deal")—are increasingly tending to follow a broad stakeholder approach alongside a "classic" investor orientation and to address a rather diffuse "interested public" as a target group for regulation. However, it often remains controversial how this change can be measured, whether it is value-relevant for companies and to what extent and for whom it can be desirable at all. In any case, the cost-benefit analyses carried out in the regulatory context are regularly very vague in terms of their benefit analysis. Work that at least makes the business benefits of these developments measurable is therefore particularly desirable.

Secondly, the particular appeal of the topic of this study also lies in the fact that the published sustainability information is not primarily based on "environmental" information, but rather on "social" information. This is a subject area that is currently still underrepresented in the overflowing ESG reporting research. At the same time, however, it is also a subject area that is particularly suitable for establishing a link to stakeholder orientation and the public value approach, and in this respect clearly demonstrates that sustainability reporting—unlike the conceptual approaches of the ISSB, for example—should go beyond the target group orientation towards investors, at least if the legitimacy of companies is to be broadly supported.

Thirdly, and particularly important from an economic perspective, the topic is not only dealt with normatively, but also analyzed empirically. As a starting point, Mr. Heroth chooses the UN Sustainable Development Goals, in which he sees a decisive driver for the legitimacy of the organization in the corporate context. The complex topic is thus made accessible to empirical research and examined using a global sample of corporate reports differentiated by industry by means of content analysis. The analysis of "Public Value Topic Salience and Sustainability Performance" is of the highest novelty content; there is practically nothing comparable in the relevant literature to date.

The implications of this work are extremely diverse: it encourages discussion and further research into the interplay between corporate reports and stakeholder value or public value and calls for the usefulness of these reports to be seen in a wider social context. This also addresses the practical relevance of the work, because as Mr. Heroth impressively demonstrates: "Corporate Sustainability Disclosure matters, also financially"! This innovative work is therefore recommended to both theorists and practitioners of corporate reporting, the (sustainable) finance sector, legislators and regulators, as well as all those who are interested in issues relating to the legitimacy of companies, their social responsibility and the UN Sustainable Development Goals.

St. Gallen
November 2024

Prof. Dr. Thomas Berndt

# Acknowledgements

I would like to take this opportunity to express my gratitude for the support I have received from many people around me during my research. Even though I cannot mention all of them in writing, I want to emphasize those who have particularly inspired and encouraged me in my work. First and foremost, my sincere gratitude goes to my supervisors, Prof. Dr. Thomas Berndt, who always had an open ear for my thoughts and ideas and, in addition, critically questioned my research in a very constructive and supportive manner and supplemented it with great ideas and approaches. And to Prof. Dr. Timo Meynhardt, who, with his wealth of experience in the field of public value, gave me helpful tips and advice for my theoretical and empirical approach in his motivating manner.

I also would like to thank my colleagues at the Institute of Financial Services Zug IFZ for supporting me on this path, and my fellow doctoral students at the University of St. Gallen, foremost Andreas Weyer and Alexander Sigg, who, during my PhD journey, shared my affiliation on sustainability reporting and stakeholder orientation and had an open ear and food for thoughts.

The most profound gratitude goes to my family, whom I dedicate this thesis. To my partner Verena Thormann and my son Eliah, who were never annoyed when I opened the laptop to recheck my data again, but on the contrary, who always supported me during this exciting journey with their affection, love, and words of support. Finally, with this acknowledgment, I also would like to thank my parents, Sigrid Heroth and Hans-Werner Heroth, and my sister Melanie Moritz, who unconditionally supported and guided me and paved the way for all the steps I have taken in my life until today.

Riehen  
January 2024

Timo Heroth

# Abstract

This doctoral thesis deals with a question that decisively influences and drives companies, regulators, banks, and investors in the 21st century. What significance do changes in social discourse, social actors, and their expectations have with regard to the legitimacy and success of companies? The aim of this dissertation is to empirically analyse the extent to which the orientation of companies with regard to their stakeholders and the inclusion of key social issues in organizational actions and corporate reporting is an influencing factor for the sustainable and financial performance of companies. In the context of increasing demands on the sustainability of companies, in addition to the more strictly defined sustainability regulations, society with its key issues, which it brings into the public discourse, is a key player with a say in the legitimization of economic activity.

The dissertation comprises two parts, the first of which is based on a theoretical framework that is generally accepted and elaborated in academia. Firstly, the development and significance of legitimacy as a concept and subsequently as a reference for organizational action must be presented. Secondly, the focus is on the characteristics of stakeholders in the corporate context, which have already been extensively researched with the help of stakeholder theory. The classification of social issues on the basis of individual human needs was defined and applied within the framework of public value theory and thus, thirdly, completes the theoretical basis of this dissertation in order to make discourse changes and important drivers for the legitimization of companies more measurable. It is time to provide empirical evidence of the importance of social stakeholders and key societal issues for the sustainable and financial value of companies.It is time to provide empirical evidence of the importance of social stakeholders and material social issues for the sustainable and financial value of companies. The second part of the thesis therefore deals with empirical evidence. It is based on a database of corporate reports from the 15 largest companies by market capitalization from six different economic sectors. The analysis period covers the years 2018–2021. In two different empirical analyses, which include both automatic and manual text analysis, the value relevance of stakeholder orientation and the materiality of social issues related to public value are investigated.

The first empirical analysis shows in detail that the orientation towards social stakeholders in sustainability reporting is positively associated with sustainable and financial performance. The second analysis provides empirical evidence that a change in the social discourse has taken place and that the orientation towards the Sustainable Development Goals (SDGs) and the related topics and values is positively associated with general sustainable performance and social sustainable performance according to the public value theory. A positive association can also be demonstrated with regard to the metric that measures alignment with the UN human rights covenant. In terms of financial performance, a positive association can be demonstrated for one SDG, SDG 4 with the topic of quality of education. In addition, the analysis of the value dimensions resting on the public value landscape according to Meynhardt (2009) shows that the moral-ethical dimension with SDGs 5 and 10, which relate to gender equality and the reduction of inequalities, has the greatest impact on both the general sustainability rating and the social pillar of the sustainability rating. In the general sustainability rating, the moral-ethical dimension is followed by the instrumental-utilitarian dimension with SDGs 4, 8 and 12, which focus on quality education, decent work and economic growth as well as responsible consumption and production. The hedonistic-aesthetic dimension and the political-social dimension also show significant results and rank third and fourth in terms of their correlation with sustainability performance. In the social pillar of the ESG rating, the moral-ethical dimension is followed by the political-social dimension, followed by the political-social dimension, the hedonistic-aesthetic dimension and the instrumental-utilitarian dimension, all of which show significant results.

# Zusammenfassung

Die vorliegende Doktorarbeit befasst sich mit einer Fragestellung, die Unternehmen, Regulatoren, Banken und Investoren im 21. Jahrhundert entscheidend beeinflusst und umtreibt. Welche Bedeutung haben gesellschaftliche Diskursveränderungen in Richtung eines nachhaltigen Wirtschaftens und soziale Akteure und deren Erwartungen bezüglich der Legitimation und des Erfolgs von Unternehmen. Ziel der vorliegenden Dissertation ist es empirisch zu analysieren, inwieweit die Orientierung von Unternehmen bezüglich Ihrer Stakeholder sowie der Einbezug von wesentlichen sozialen Themen in das organisationale Handeln und die Unternehmensberichterstattung einen Einflussfaktor für die nachhaltige und finanzielle Performance von Unternehmen darstellt. Im Kontext der steigenden Anforderungen an die Nachhaltigkeit von Unternehmen ist neben der strenger definierten Nachhaltigkeitsregulatorik, die Gesellschaft mit Ihren wesentlichen Themen, die sie in den öffentlichen Diskurs einbringt, ein wesentlicher Akteur mit Mitspracherecht bezüglich der Legitimation von wirtschaftlichem Handeln. Die Dissertation umfasst zwei Teile, wobei der erste Teil auf einem in der Wissenschaft allgemein akzeptierten und ausgearbeiteten Theoriegerüst fusst. Als Erstes gilt es die Entwicklung und Bedeutung der Legitimation als Konzept und in der Folge als Referenz von organisationalem Handeln darzustellen. Zweitens stehen die Charakteristiken von Stakeholdern im Unternehmenskontext im Vordergrund, die mit Hilfe der Stakeholder Theorie bereits umfassend erforscht sind. Die Klassifizierung von sozialen Themen auf Basis individueller menschlicher Bedürfnisse wurde im Rahmen der Public Value Theorie definiert und angewendet und komplettiert damit, drittens, die theoretische Grundlage der vorliegenden Dissertation, um Diskursveränderungen und wichtige Treiber zur Legitimation von Unternehmen messbarer zu machen. Es ist es an der Zeit empirische Belege für die Bedeutung sozialer Stakeholder und wesentlicher gesellschaftlicher Themen für den nachhaltigen und finanziellen Wert von

Unternehmen zu liefern. Der zweite Teil der Arbeit befasst sich folglich mit der Empirie. Grundlage stellt dabei eine Datenbasis von Unternehmensberichten der nach Marktkapitalisierung 15 grössten Unternehmen aus sechs verschiedenen Wirtschaftssektoren dar. Der Analysezeitraum umfasst die Jahre 2018–2021. In zwei unterschiedlichen empirischen Analysen, die sowohl eine automatische als auch manuelle Textanalyse umfassen, wird die Wertrelevanz der Stakeholderorientierung sowie die Wesentlichkeit von mit Public Value in Bezug stehenden sozialen Themen erforscht. Die erste empirische Analyse zeigt im Detail auf, dass die Orientierung gegenüber sozialen Stakeholdern in der Nachhaltigkeitsberichterstattung positiv mit nachhaltiger und finanzieller Perfomance assoziiert ist. Die zweite Analyse liefert einen empirischen Nachweis, dass eine Veränderung des gesellschaftlichen Diskurses stattgefunden hat und die Ausrichtung gegenüber den Sustainable Development Goals (SDGs) und den damit in Bezug stehenden Themen und Werten nach der Public Value Theorie, positiv mit der generellen nachhaltigen Performance und der sozialen nachhaltigen Performance assoziiert ist. Eine possitive Assoziation lässt sich ebenfalls bezüglich der Metrik, die die Ausrichtung gegenüber der UN Menschenrechtskovention misst, nachweisen. Bei der finanziellen Performance lässt sich eine positive Assoziation bei einem SDG, dem Ziel Nummer 4 mit dem Thema der Qualität der Bildung, nachweisen. Zudem zeigt sich bei der Analyse der Wertdimensionen anhand der public value landscape nach Meynhardt (2009), dass die moralisch-ethische Dimension mit den SDGs 5 und 10, die sich auf die Gleichstellung der Geschlechter und die Verringerung von Ungleichheiten beziehen, sowohl die größten Auswirkungen auf das allgemeine Nachhaltigkeitsrating, als auch auf die soziale Säule des Nachhaltingkeitsratings hat. Beim allgemeinen Nachhaltigkeitsrating folgt auf die moralische-ethische Dimension die instrumentell-utilaristische Dimension mit den SDGs 4, 8 und 12, die sich auf qualitativ hochwertige Bildung, menschenwürdige Arbeit und Wirtschaftswachstum sowie auf verantwortungsvollen Konsum und Produktion konzentrieren. Die hedonistisch-ästhetische Dimension und die politisch-soziale Dimension zeigen ebenfalls signifikante Ergebnisse und rangieren hinsichtlich ihrer Korrelation mit der Nachhaltigkeitsleistung an dritter und vierter Stelle. Bei der sozialen Säule des ESG Ratings folgt auf die moralisch-ethische Dimension die politisch-soziale Dimension gefolgt von der politischsozialen Dimension, der hedonistisch-ästhetischen Dimension, und der instrumentell-utilaristischen Dimension, die alle signifikante Ergebnisse aufzeigen.

# Contents

# Abbreviations

| | |
|---|---|
| ASB | Auditing Standards Board |
| B.C.E. | Before the Common Era |
| BERT | Bidirectional Encoder Representations from Transformers |
| BICS | Bloomberg Industry Classification Standard |
| CERDS | China Enterprise Reform and Development Society |
| CHF | Swiss Francs |
| CO | Swiss Code of Obligations |
| CSR | Corporate Social Responsibility |
| CSRC | China Securities Regulatory Commission |
| CSRD | Corporate Social Responsibility Directive |
| EBIT | Earnings Before Interest and Taxes |
| EBITDA | Earnings Before Interest, Taxes, Depreciation, and Amortization |
| EFRAG | European Financial Reporting Advisory Group |
| EPS | Earnings per Share |
| ESG | Environmental, Social, Governance |
| ESRS | European Sustainability Reporting Standards |
| EU | European Union |
| FSB | Financial Stability Board |
| GDP | Gross Domestic Product |
| GRI | Global Reporting Initiative |
| IASB | International Accounting Standards Board |
| IFRS | International Financial Reporting Standards |
| IRRC | International Integrated Reporting Council |
| ISSB | International Sustainability Standards Board |
| LDA | Latent-Dirichlet-Allocation |
| LLM | Large Language Model |
| MDandA | Management's Discussion and Analysis |

| | |
|---|---|
| NFRD | Non-Financial Reporting Directive |
| NGO | Non-Governmental Organization |
| PVTS | Public Value Topic Salience |
| RBI | Responsible Business Initiative |
| ROA | Return on Assets |
| ROCE | Return on Capital Employed |
| SASB | Sustainability Accounting Standards Board |
| SDG | Sustainable Development Goals |
| SEC | United States Securities and Exchange Commission |
| SFDR | Sustainable Finance Disclosure Regulation |
| SGMM | St. Gallen Management Model |
| SOV | Share of Voice |
| SR | Social Responsibility |
| TCFD | Task Force on Climate-Related Financial Disclosures |
| UN | United Nations |
| UNGC | United Nations Global Compact |
| U.S. | United States |
| USA | United States of America |
| VIF | Variance Inflation Factors |
| WACC | Weighted Average Cost of Capital |
| WBCSD | World Business Council for Sustainable Development |

# List of Figures

# List of Tables

# List of Equations

# 1 Introduction

The world is in a state of flux. So are the dynamics and subjects shaping the relationship between business organizations and their external environment. Apart from a fluidity of opinions and stories stand societal issues and values that become essential to the corporate agenda. This development is not only a matter of reputation but a crucial component for the legitimacy and survival of companies.

This dissertation refers to a primary driving force for reshaping financial markets and the business world in the 21st century, namely the topic of sustainability, its related concerns, and disclosure. Sustainability represents an influencing factor in organizational activities and creates new business models, standards, and ratings that question the current understanding of business.

What drives the corporate agenda to focus more on sustainability and social issues in recent years? By facing the grand challenges[1], the world is in confrontation with tremendous problems such as climate change and related human rights issues. Companies have to fulfill mandatory sustainability regulations being on the rise in various countries and geographies, such as the Corporate Sustainability Disclosure Regulation (CSRD) in the European Union (EU). However, besides stricter laws and directives, companies are also facing a discursive shift that influences how actors, such as corporate stakeholders grasp and relate to sustainable development in their communication,[2] and what the public expects from companies to keep their "social license to operate"[3].

[1] For further research on the grand challenges, see George et al. (2016).

[2] Biermann et al. (2022), p. 795.

[3] Boutilier and Thomson (2011), p. 4.



T. Heroth, *Corporate Sustainability Disclosure Matters*,
https://doi.org/10.1007/978-3-658-47653-3_1

This development raises the question of how companies can build and maintain a "contextual sensitivity"[4] for public issues and their connected stakeholders and, with the development towards higher expectations for business organizations of being sustainable, if and how such a discursive shift is linked with corporate value. Sustainability as a critical component of corporate actions and its risk criteria comes with the questioning of existing models that define value in a market economy based on neoclassical theories. Calling for more recent theoretical approaches that consider the relevance of corporate stakeholders and social issues, theories such as stakeholder theory and public value theory enable the exploration of the constructs that are in the focus of this dissertation, namely the value relevance of stakeholder orientation and public value topic salience.

With the help of a growing data pool of corporate reports that includes an increasing number of content related to sustainability, this dissertation draws on the analysis of textual information being part of corporate reports to unpuzzle corporate stakeholder orientation and the issues that drive this relationship. The following section emphasizes the motivation for the research in focus.

## 1.1 Research Motivation

The dissertation focuses on a field of tension in which companies, due to stakeholder expectations, are required to position themselves towards their stakeholders and relevant social issues in times when sustainability regulations slowly reach greater maturity, undeniable environmental and social problems become more severe, and new business models evolve that address and support the basic needs of societies and its people.

The expectations addressed to businesses to make more significant shifts toward sustainable actions and towards its disclosure are manifold. Exploring how companies address these expectations and the stakeholders behind provokes the motivation of the author of this dissertation to understand its value relevance and impact on organizational legitimacy. On the one hand, these shifts can result from regulations or reputational reasons linked to a fear of missing transparency expectations that could lead to a lack of trust by internal and external stakeholders. On the other hand, perceptions of what value is and what is relevant for its accumulation can also result from a discourse taking place in society that further shapes the understanding of organizational legitimacy.

[4] Høvring et al. (2018), p. 641.

An additional motivation for this dissertation is to gain further empirical evidence in a widely discussed but marginally explored field. Although the sustainability reporting domain is rich in information that increases complexity, more granular explorations with verifiable and valid empirical evidence focusing on more detailed constructs, such as stakeholder orientation and public value issues that provide more clarity, still need to be contributed.

The so far unvalued public discourse triggers the author of this dissertation to make discursive shifts and its impact on organizational legitimacy measurable. With the help of widely accepted theories, such as stakeholder theory and public value theory, and the existing knowledge about the characteristics of organizational legitimacy, it is the right time to generate further empirical evidence on the constructs in focus.

These circumstances motivate to dive deeper into the rich data pool of corporate sustainable reports to unpuzzle the value relevance of stakeholders and its salient issues and to enable further research in this field. Besides, the benefits of this research for practitioners such as standard setters and rating agencies in the development of evolving sustainability disclosure standards and sustainability ratings are apparent. Apart from researchers, regulators, standard setters, rating agencies, investors, and banks, companies can benefit from this dissertation that deals with the relevance of the necessary ingredients of corporate disclosure, helping to guide organizations in their exchange with the public space and to manage their salient issues.

## 1.2 Research Background

Though still in formation, current trends towards stricter, more binding, and explicit regulations on how companies disclose exposures to sustainability issues and opportunities significantly contribute to this research. With increasing clarity on what to report and how to report, classification frameworks and taxonomies such as the materiality map of the Sustainability Accounting Standards Board (SASB), the standards of the Global Reporting Initiative (GRI), and the Sustainable Development Goals (SDGs) with its approach to double materiality, help to deliver a structure for the analysis of sustainability topics and public value related issues. Thus, it enables the analysis of the relevance and impact of concrete topics going beyond the very high-level perspective on environmental, social, and governance (ESG) concerns that is the subject of various publications in the sustainable business and reporting domain. Stakeholders, with its most salient issues being addressed in corporate sustainability reports, provide the data basis for this exploration.

At the same time, aligning corporate actions with stakeholder expectations becomes a risk and value criterion for companies. Agreement exists that a misalignment of expectations between a company and its financial stakeholders could be a risky and value-destructive endeavor for a company. As the St. Gallen Management Model (SGMM) postulates, corporate value creation is largely influenced by expectations that are "the condensed, anticipated future"[5]. A company acting in contradiction or conformity with those expectations is subject to evaluations. Public value theory agrees by emphasizing that "values are the result of evaluations, influencing future and present behavior"[6].

However, companies' external environment is not only driven by financial stakeholders and their expectations but, to an increasing extent, by the public and its societal expectations. Therefore, this dissertation focuses on analyzing the discursive shift toward societal stakeholders and social issues to analyze the value relevance and potential power to become a crucial component of organizational legitimacy. Based on Suchman's (1995) understanding of organizational legitimacy, this dissertation puts a lens on a reference point for corporate actions that are desired by society and appropriate in the value dimensions and norms that are tackled.[7]

Understanding value as a notion triggered by relationships and issue-related actions of companies leads to the concept of public value. Issues related to individuals' basic needs are part of the topical canon of public value. Destroying public value by corporate actions means harming the basic needs of individuals. Approaching the concept of public value from a sustainability reporting perspective sheds light on the orientation of companies towards their stakeholders and the value that public value and its related issues contribute to a company's sustainable and financial performance.

Exploring and measuring value-relevance requires a solid definition, what underlying concepts drive its creation and growth, and, most importantly, empirical data to gather relevant information. This links theory with empirical insights and brings sustainability frameworks such as the SDGs and their subtopics into place to enable a goal and issue-oriented analysis.

The illustration of the research gaps in the following chapter shows that business and management research asks for more research projects that unpuzzle concepts such as the measurability of public value and organizational legitimacy with a focus on sustainability reporting. With the help of recent contributions in stakeholder theory, public value theory, and sustainability standards and frameworks, the author

[5] Rüegg-Stürm and Grand (2019), p. 137.

[6] Meynhardt (2009), p. 199.

[7] see Suchman (1995), p. 574.

of this dissertation aims to explore the value relevance of stakeholder orientation and public value topic salience for corporate value creation and to close existing research gaps.

## 1.3 Research Gaps and Problem Statement

Developing from concepts on the obligations of business organizations with regard to their external environment that some authors, already in the 1980s, summarized as stakes that go beyond ownership,[8] stakeholder theory gained attention in recent years. In this context, publications by authors such as Freeman (1984) defined the role of stakeholders for business organizations as influential. Stakeholder theory distances itself from positions that limit the view of organizations to its shareholders and counteract statements that label the integration of stakeholders as mere stakeholderism and as an illusionary promise.[9]

With the development towards a more significant influence of sustainability on business organizations, the impact of corporate stakeholders and its related issues gained further attention from practitioners and researchers focusing on subjects such as stakeholder engagement, stakeholder dialogue, and stakeholder orientation.[10] This development enabled the consideration of corporate sustainability disclosure as a relevant source of information for analyzing the meaning and role of stakeholders in the corporate context.

However, besides publications that included the role of single stakeholders such as community, shareholders, customers, suppliers, and the government and their impact on the quality of stakeholder engagement processes,[11] research on the value relevance of addressing stakeholders such as social stakeholders in corporate reporting is lacking. This first research gap shall be closed by the first empirical investigation of this dissertation.

Therefore, a rule-based text analysis with the help of stakeholder dictionaries that draws on existing research publications and sustainability standards, such as the GRI Disclosure Directive 2-29 with its approach towards stakeholder engagement, allows for a detailed analysis of stakeholder orientation. Sustainability ratings and

[8] see Jones (1980), pp. 59-60.

[9] cf. Bebchuk and Tallarita (2020), Bebchuk et al. (2020).

[10] Bettinazzi and Zollo (2017); Bettinazzi and Feldman (2021); Harrison et al. (2010); Zollo et al. (2009).

[11] Stocker et al. (2020).

financial performance metrics further help to explore the value relevance of the topic in focus.

The second research gap concerns the underlying issues driving a corporate stakeholder relationship. Bringing in the public as a relevant point of reference raises the question of how salient topics and issues that are part of the relationship between companies and society can represent a measurable construct. This leads to the consideration of public value theory since it describes value as the quality and "the result of a relationship"[12]. In public value theory, basic individual needs are necessary to guarantee quality of life and well-being. The corporate perspective relates to public value by asking how corporate actions positively or negatively contribute to fulfilling basic needs. Public value is present and measurable if corporate actions that relate to basic needs and their related value dimensions, from the perspective of society, do not reveal a discrepancy between the experience of an action and what a person classifies as appropriate.[13]

Therefore, measuring public value relates to corporate actions and societal discursive shifts. It refers to the contribution of public value to the maintenance, creation, and potential loss of organizational legitimacy. It asks for a framework that enables the classification of individual basic needs and a reference for what society classifies as appropriate actions and necessary to address in corporate disclosure.

Research on public value beyond public management has mainly provided valuable theoretical contributions.[14] Empirical-oriented publications focusing on organizational public value deal with topics such as employee life satisfaction[15] and ethical leadership.[16] Existing research also includes a view on corporate disclosure and non-financial reporting.[17] However, the lack of empirical investigations exploring organizational public value in a business context is generally criticized.[18]

Thus, focusing on corporate sustainability reporting by drawing from the vast literature on public value theory and implementing sustainable development frameworks such as the SDGs that allow measuring public value and its impact on organizational legitimacy shall close this research gap. Consequently, as the following section shows, the goal of this dissertation is to prove empirical evidence on the dis-

[12] Meynhardt (2009), p. 198.

[13] see Meynhardt (2009); Meynhardt (2015).

[14] Benington and Moore (2011); Meynhardt et al. (2017); Meynhardt and Bäro (2019).

[15] Meynhardt et al. (2020).

[16] Sami et al. (2016).

[17] Meynhardt and Bäro (2019).

[18] Hartley et al. (2017).

cursive shift that happened in society about the role organizations play in fulfilling basic needs and in taking a necessary responsibility towards related social issues.

Figure 1.1 shows how stakeholder and public value theory are linked with the concept of corporate value. The indicator share of voice (SOV) builds the evaluation criteria for stakeholder orientation. The indicator public value topic salience (PVTS) allows for evaluating basic needs and their most salient topics. By testing for sustainability and financial performance with the help of ESG ratings and the cost of capital of companies, the goal is to show that a discursive shift took place in society, affecting the understanding of organizational legitimacy.

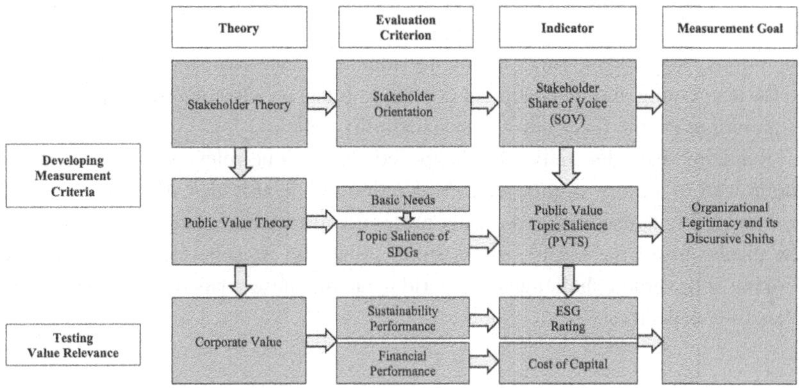


**Fig. 1.1** Stakeholder Theory, Public Value Theory, and Corporate Value with its Evaluation Criteria, Indicators, and its Measurement Goal

## 1.4 Research Objective

By linking existing findings on organizational legitimacy, stakeholder theory, and public value theory, further research might draw from this link to generate findings that contribute to and enrich existing theories. However, this dissertation's primary and first goal is an empirical one. It is finding evidence for the measurability of organizational legitimacy, and the role social stakeholders and issues play in that context. A proof shall be delivered that shows that a discursive shift happened towards the necessity of companies to be ahead of societal discourse and of issues that drive public value. The sustainable and financial value relevance of stakeholder orientation and public value topic salience would provide this proof.

A second goal is to find evidence if and how specific social and financial stakeholders and the most salient social issues are associated with sustainability and financial performance. With this evidence, the third goal of this dissertation is to provide readers, be it out of a practical or theoretically induced interest, an approach that motivates readers to focus on social stakeholders and their expectations. Focusing on the issues around social stakeholders shall further support regulators, standard setters, rating agencies, banks, investors, and companies to gain a deeper understanding of the meaning and relevance of public value and its salient topics and how these topics can be better implemented in sustainability standards and better addresses in sustainability disclosure and corporate actions.

For researchers, the research approach of this dissertation aims to result in greater motivation for including existing frameworks, such as the SDGs, in empirical explorations. At the same time, its link to value relevance shall motivate researchers in the accounting and management domain to focus on sustainability reporting and explorations on the materiality of sustainability topics.

Realizing that a discursive shift happened, that is value-relevant for organizations might lead to a more significant consideration of social issues affecting corporate actions. It might also lead to a greater professionalization of companies concerning the disclosure of the social sustainability pillar and its subtopics. Consequently, looking at the grand challenges the world is facing, these goals fulfill a wish for the planet we depend on.

## 1.5 Research Question and Hypotheses

The following research question aims to reach the objectives mentioned above and to close the research gaps:

*How are stakeholder orientation (SOV) and public value topic salience (PVTS) associated with sustainable and financial performance?*

As further elaborated in Sect. 4.2, which focuses on the managerial perspectives and the classification of stakeholders, the hypotheses differentiate between two groups of stakeholders, social and financial stakeholders, to be able to isolate and test the impact of stakeholder orientation and public value topic salience on the dependent variables in focus.

The research question leads to constructing the following hypotheses that focus on analyzing stakeholder orientation and its association with sustainability performance.

*H1a Social stakeholder orientation is positively associated with sustainability performance.*

*H1b Financial stakeholder orientation is negatively associated with sustainability performance.*

Linking stakeholder orientation with the value relevance focusing on financial performance leads to the development of the following hypotheses:

*H2a Social stakeholder orientation is negatively associated with the cost of capital.*

*H2b Financial stakeholder orientation is positively associated with the cost of capital.*

Shifting the focus on the exploration of public value topic salience, its association with sustainability and financial performance, and the intention to answer questions related to the second research gap leads to hypotheses H3 and H4:

*H3 Public value topic salience is positively associated with sustainability performance.*

*H4 Public value topic salience is negatively associated with the cost of capital.*

## 1.6 Methodology and Scope

Text analytic studies provide valuable insights into sustainability reporting because the basis of such information is data that is available in alternative and unstructured text data format.[19] However, there is a lack of existent studies that embrace text analytics to analyze stakeholder orientation as a crucial component of corporate reporting. This asks for further empirical research. Therefore, this dissertation embraces text analytics as a method and a tool to explore the stakeholder orientation of global companies with the help of a rule-based approach to gain insights into the managerial perspective of corporate stakeholders.

To construct the share of voices in the first stage, a stakeholder dictionary draws from existing publications and contributions provided by standard setters such as GRI to help classify stakeholders. In the second stage, the SOVs are associated with the Refinitiv and Bloomberg sustainability ratings and the respective companies' capital costs. A panel regression that relates the scores for stakeholder orientation and its SOVs with relevant sustainability and financial performance metrics leads

[19] see Baier et al. (2020); Barman (2018); Capelle-Blancard and Petit (2019), Goloshchapova et al. (2019), Jaworska and Nanda (2016).

to the establishment of empirical evidence to shed light on the value relevance of stakeholder orientation in corporate reporting.

Exploring public value topic salience (PVTS) rests on the SDG framework and its link to the public value landscape by Meynhardt (2009). The number of total SDGs mentioned and the concrete Sustainable Development Goal in focus build the public value topic salience metric for the analysis of the second empirical part of this dissertation. Running panel regressions that link the PVTS metric with the sustainability ratings provided by Bloomberg and Refinitiv and the cost of capital of the individual companies results in the detection of the sustainable and financial value relevance of public value topic salience.

The scope of this research does not include a perspective on sustainable finance or impact investing. Instead, the research lens lies on sustainability disclosure and the constructs in focus. Providing a holistic picture of further existing theories focusing on value is beyond the scope of this dissertation. Furthermore, the concept of sustainability is limited to the perspective of organizations and their reporting activities rather than to more general and related perceptions of the concept stemming from fields such as sociology.

The author makes no claims to completeness concerning this dissertation's different views and research approaches that are taken into account. Though a broad literature review of international literature is part of this dissertation, from a cultural and geographical perspective, concept developments, theories, and content might be dominantly influenced by German and English-speaking literature and its related views. Limitations related to methodology and the empirical investigation of this dissertation are part of the empirical chapter.

## 1.7 Structure of the Thesis

This dissertation consists of 14 chapters. Chapter 1 represents the introduction of this thesis, including summarized information on its theoretical background, the motivation for the topic, the existing research gaps and objectives, and its methodological and structural approach.

Part A of the dissertation, starting with Chap. 2, contains the why behind the included theories and remarks on the theoretical development of organizational legitimacy. Links to stakeholder and public value theory represent suitable contributions of researchers helping to introduce and illustrate the central theme of the research endeavor.

Chapter 3 covers the theoretical development and foundations, starting with the general and historical background of organizational legitimacy while further elaborating on its development from a closed system to an open system view that ignited the consideration of the organizational environment. With its connection to the St. Gallen Management Model that entails an open system view, the chapter further deals with organizational legitimacy as a construct of knowledge and communication. Being shaped by society and issues that are part of societal discourse, the concept of organizational legitimacy demands a dynamic view of a changing organizational environment.

Chapter 4 focuses on stakeholder theory, its scientific contestation, recent publications, and its views and classifications. This chapter puts the external environment, its players, and the issues that evolve these actors into the context of organizational legitimacy and public value. It also sets the seeds for the empirical investigation, taking a closer look at the categorization of social and financial stakeholders.

Chapter 5 concentrates on public value theory with a focus on social issues. The chapter explains why public value describes the quality of relationships and why basic needs and the related values on the public value landscape by Meynhardt (2009) serve as a reference point for organizational legitimacy and a suitable framework for measuring organizational legitimacy. This chapter ends with explaining the link between societal discourse and public value. Chap. 6 summarizes the theoretical contributions and relevant findings of this dissertation.

Chapter 7 invites the reader to dive deeper into sustainability, its historical foundations, organizational implications, and significance for corporate disclosure. The chapter outlines the regulatory environment in the EU and other geographies and countries. The regulatory framework leads to understanding the concept of materiality, with its different approaches pursued globally. The landscape of reporting standards is a further part of this chapter, finishing with the SDGs as one approach towards double materiality. Chap. 8 highlights the critical aspects of sustainability and materiality.

Part B of this dissertation, the empirical part, starts with Chap. 9, which deals with the research gap, followed by Chap. 10, narrowing down the research questions and, Chap. 11, describing the data sample for the analysis. Chap. 12 sets the boundaries of the research in focus by emphasizing its limitations related to the data sample, its technical or empirical approach, its general content, or cultural reasons.

Chapter 13 introduces content analysis with its reference to Mayring (2014). It illustrates the steps in more detail that Mayring (2014) defined for analyzing textual content with the help of a structured approach that includes the definition of the material in focus, the document origins, its quality criteria, and its formal characteristics. The empirical investigation is divided into two parts. The first analysis uncovers stakeholder orientation with the help of descriptive and inferential statistics. The second exploration reveals salient topics of public value. Both investigations are tested on their value relevance in terms of sustainability and financial performance. With the help of the structure provided by Mayring (2014), the empirical investigations are rechecked on their alignment with existing theoretical findings and with an interpretation of its results, including its implications for regulators, standard setters, rating agencies, investors, banks, and companies. This dissertation ends with a general conclusion of the empirical findings and a discussion and outlook focusing on this topic's future development and potential (Chapter 14). The following Fig. 1.2 visualizes the structure of this dissertation.

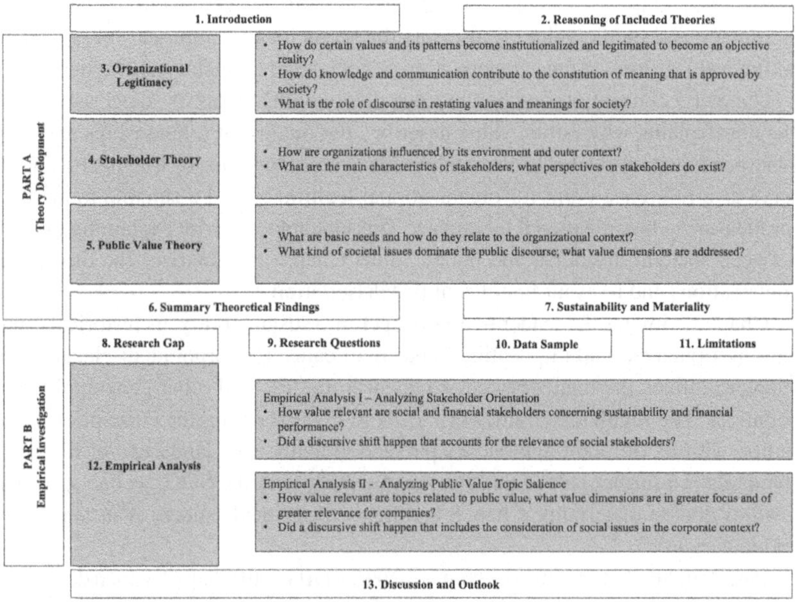


**Fig. 1.2** Structural Overview of the Dissertation

## 1.8 Research Design

This dissertation focuses on the empirical investigation of stakeholder orientation and public value topic salience. Therefore, it follows a deductive approach that draws on existing theories and theoretical contributions, namely on organizational legitimacy (Chap. 3), stakeholder theory (Chap. 4), and public value theory (Chap. 5). This order suits the transition to the empirical investigation and the deduction of empirical findings since both empirical constructs, stakeholder orientation, and public value topic salience, are considered from the perspective of organizational legitimacy.

Stakeholder theory helps to identify the main stakeholders and their characteristics for measuring stakeholder orientation and their stake in organizational legitimacy. Public value theory allows to measure organizational legitimacy with the help of its value dimensions and related basic needs that are subject to the SDG framework. Therefore, the hypotheses of this dissertation are derived from the mentioned theoretical foundation and further tested with the help of a panel data regression analysis (Chap. 12). To better classify and appreciate this topic's importance in context of corporate sustainability reporting, a single chapter covers the development of sustainability disclosure, its regulatory background, and the role of approaches toward single and double materiality (Chap. 7).

The independent variables that represent stakeholder orientation and public value topic salience are generated with the help of a rule- and dictionary-based approach. Exploring stakeholder orientation draws on text analysis with the help of coded dictionaries, including the main corporate stakeholders, their synonyms, and related words. Public value topic salience analysis rests on a manual analysis of content reported in corporate disclosure. The empirical steps of this dissertation focus on qualitative information in the form of text that is transformed into quantitative information. With the help of the SOVs of stakeholder orientation that consider its mention frequency regarding the total words mentioned in a document and with the help of the sum and single SDGs mentioned in the corporate reports of the data sample, the results for the panel regressions are retrieved. Testing for value relevance and impact on organizational legitimacy is empirically realized with the help of the dependent variables that relate to sustainability and financial performance. Sustainability performance covers the sustainability scores of the well-accepted providers Bloomberg and Refinitiv. Bloomberg also provides the financial data of this dissertation, such as the companies' cost of capital, including its cost of debt and cost of equity.

This innovative approach that includes text analytics and manual exploration enables the measurement of organizational legitimacy to explore the value relevance of societal discourse, social issues, and stakeholders being part of the corporate environment.

# Part I

# Theory Development

# Reasoning of Included Theories 2

This dissertation refers to three theories that build its theoretical foundation: organizational legitimacy as part of legitimacy theory, stakeholder theory, and public value theory. Academic research may lack a consistent focus if there is more than one theory included. However, since this dissertation is a monographic illustration of two more extensive constructs, which are stakeholder orientation and public value topic salience, the author considers the consideration of the mentioned theories as a suitable and necessary approach.

As the subsequent chapters show, stakeholder theory and public value theory build on the concept of organizational legitimacy. Researchers enriched the existing knowledge with crucial contributions, such as Suchman (1995), who defined legitimacy in a way that reduced the complexity of the term and made its analysis, including its empirical investigation, more tangible. However, research also showed that developments such as discursive shifts in society challenge the current definition of organizational legitimacy, asking how to give organizations' outer context, dynamics, and stakeholders a more significant stake. Being shaped by knowledge and communication, organizational legitimacy builds upon values and habits that become institutionalized, legitimated, and accepted by society. Nevertheless, while becoming an objective reality, organizational legitimacy challenges organizations and their actions. Public acceptance turns into a reference point for organizations influenced by societal discourse and public will. Therefore, organizational legitimacy institutionalizes and legitimates the meaning of values, their related patterns, and issues that address and challenge organizations from their outer context.

Considering the outer context of organizations leads to stakeholder theory that describes the players of the corporate environment, its values, expectations, and the power and urgency that these players bring into the organizational arena. Organizations are nowadays challenged to incorporate the organizational environment into their strategies and operational business. On the one hand, in a more pessimistic



T. Heroth, *Corporate Sustainability Disclosure Matters*,
https://doi.org/10.1007/978-3-658-47653-3_2

view, there is a danger that organizations lose touch with their environment, which leads to questioning organizational legitimacy and, consequently, to critical scenarios of organizational survival. On the other hand, stakeholder theory shows an exchange with stakeholders, and aligning with their issues represents an explicit business task.

Corporate actions focusing only on symbolic and visible signs of stakeholder engagement will lack organizational legitimacy.[1] Symbolic actions lack moral legitimacy, leading to "moral indignation and further reduce public acceptance"[2]. Ashforth and Gibbs (1990) emphasized that it is the authentic, salient, and meaningful practices of businesses that matter and that need to be aligned with social expectations while business operations are further running. Consequently, there needs to be more research related to stakeholder theory that includes an empirical approach to explore the value relevance of organizational stakeholders and discursive shifts in society.

This dissertation embraces stakeholder theory since it acknowledges the role of relevant stakeholders as part of the outer context of organizations. It considers stakeholders such as society with expectations assumed to be relevant for organizations to establish and maintain organizational legitimacy.

Public value theory and its former contributions by researchers enable and support the measurement and identification of the value relevance of organizational legitimacy. Public value builds upon individual basic needs that become a reference point for the values of the public. For organizational legitimacy, values and habits are crucial for becoming a part of an objective reality that society shapes and accepts. By unpuzzling these values and making them measurable with the help of value dimensions on the public value landscape (Meynhardt 2009), public value theory reveals what issues are a relevant part of the societal discourse and its discursive shifts.

In sum, organizational legitimacy builds a basis by explaining the concept of discourse and answering questions such as how values become institutionalized and legitimated to become objective reality. It further enlightens the reader on how meaning for society is constructed with the help of knowledge and communication. Stakeholder theory tackles questions on how the outer context and its players influence organizations and how characteristics such as urgency and the power of stakeholders shape organizational legitimacy. It contributes to answering questions on the value relevance of stakeholder orientation with the help of empirical analysis that builds upon the existing theoretical findings. Public value theory brings trans-

[1] Ashforth and Gibbs (1990), p. 181., as cited in Palazzo and Scherer (2006), p. 79.

[2] Palazzo and Scherer (2006), p. 79; Ashforth and Gibbs (1990).

parency into what is worth for society and enables the measurement of its sustainable and financial value relevance with the help of its value dimensions. It provides a structure with the help of the public value landscape and allows the exploration of societal discourse and the discursive shifts of its issues.

All three theories focus on values and issues that become part of the public discourse. They establish the theoretical foundation for this dissertation by building on each other. By explaining reasons and providing a structure, the mentioned theories enable empirical research in the field of sustainability reporting, which facilitates the exploration of the value relevance of stakeholder orientation and public value topic salience. Therefore, the following chapters and sections illustrate the origins, developments, and characteristics of the theoretical pillars of this dissertation in more detail.

# 3 Organizational Legitimacy—Its Foundations and Current Impact

Legitimacy can represent a driving force for corporate actions and a company's success and survival.[1] However, what stands behind this term, and how could legitimacy become such an influential factor in corporate actions?

According to Suchman (1995) and his well-accepted definition, legitimacy describes a "generalized perception or assumption that the actions of an entity are desirable, proper, or appropriate within some socially constructed system of norms, value, beliefs, and definitions"[2]. The definition's complexity describes the challenges of analyzing legitimacy as a construct. Suchman (1995) contributed a "vastly expanded theoretical apparatus"[3] to the research of organizational legitimacy. Different theoretical perspectives on organizational legitimacy exist, relating to theories, such as resource dependence theory,[4] institutional theory,[5] and theories of organizational ecology.[6] This is made even more complex because legitimacy is a construct in flux that depends on current perceptions and assumptions of social actors that do not change daily but are subject to discursive shifts being influenced by global and local "events" and issues that shape narratives and convictions of societies. Imerman (2018) emphasized the construct's dynamics and fluidity by speaking of legitimacy cycles.[7]

However, to narrow down the construct of legitimacy, Suchman (1995) divided organizational legitimacy into three different forms. First, pragmatic legitimacy

[1] Meyer and Rowan (1977), p. 353.

[2] Suchman (1995), p. 574.

[3] Suchman (1995), p. 571.

[4] Pfeffer and Salancik (1978).

[5] Meyer and Rowan (1977).

[6] Carroll and Hannan (1989).

[7] Imerman (2018), p. 75.



T. Heroth, *Corporate Sustainability Disclosure Matters*,
https://doi.org/10.1007/978-3-658-47653-3_3

deals with the self-interest that an audience and its individuals possess. It further addresses the well-being that might be affected by organizational action. This relates to the broader environment of organizations and their interactions with an organization. In contrast, a beneficial and materialistic exchange relationship between the organization and the outer context is the focal point. At the same time, a scrutinization of organizational actions by the audience can be part of that form.[8]

The second form, cognitive legitimacy, covers a form that is not related to the beneficial interest of an audience but more to an inevitable, subconscious acceptance of organizational actions.

The third form is not cognitive and is not related to interest. It pertains to normative evaluations of organizational actions. It relates to establishing societal welfare and "rests not on judgments about whether a given activity benefits the evaluator, but rather on whether the activity is 'the right thing to do' "[9]. What is evaluated by the moral form of legitimacy are either techniques and procedures, outputs and consequences, or categories and structures.[10] It is such the moral form of legitimacy that is the focus of this dissertation since it results from "explicit public discussion"[11] that organizations have to take part in to gain legitimacy. The goals that an organization strives for and the way organizations pursue these goals have to be acceptable from a societal standpoint.[12]

Suchman (1995) further highlighted two approaches for organizations to deal with legitimacy. First, the institutional approach, and second, the strategic approach. While the institutional approach covers organizations' compliance with societal norms, beliefs, and values,[13] the strategic approach pertains to organizational actions by which legitimacy can be controlled and steered.[14]

Diving deeper into the meaning of organizational legitimacy in the current organizational context requires the consideration of its theoretic and historical foundations that contributed to the development of what legitimacy is today. Therefore, the following section explores the roots of legitimacy in sociology and organizational theory, explaining what led to a closed system view that concentrates on the inside of organizational boundaries to an open system view that allows the external environment to become an influential player for organizational behavior.

---

[8] Suchman (1995), p. 578.

[9] Suchman (1995), p. 579.

[10] Suchman (1995).

[11] Suchman (1995), p. 585.

[12] see Ashforth and Gibbs (1990), p. 177.; Palazzo and Scherer (2006), p. 73.

[13] see DiMaggio and Powell (1983); Suchman (1995).

[14] see Ashforth and Gibbs (1990); Suchman (1995).

## 3.1 The Roots of Organizational Legitimacy—A Closed System View

In times of great changes that took place in the world during the Industrial Revolution in the late 18th and early 19th century, large organizations started to emerge with the overarching goal of increasing the efficiency of their production processes. At that time, the lens was on internal processes within the corporate boundaries of organizations. As "rational systems"[15], the concentration was on how the highest output can be reached with the lowest material input.[16]

Compared to other organizational theories and models that developed in the 20th century, such as the goal-oriented model[17] that in a simple definition focuses more on task completion and outcome assessments, the system-oriented view concentrates more on the process components of organizational performance, including the consideration of legal and moral organizational activities.[18] To allow reflections on the systems-oriented view, organizations as such are complex systems,[19] while a system is defined as an independent entity with interactional patterns between its elements.[20] Though outcomes also play a role in system-oriented models of organizations, as early proponents such as Katz and Kahn (1966) and Yuchtman and Seashore (1967) already acknowledged,[21] the focus lay more on the organizational interactions and transactions with the organizational environment that at least in the open system view came to the surface.[22] As Selznick argued in 1948, "the security of the organization as a whole is about social forces in its environment"[23], which is important for "the maintenance of the system as a generic need"[24].

However, the integration of the external environment of organizations was still to come. This is why research and conceptions of organizations under the natural and rational system models are summarized under the closed system model.[25] Industrial societies developed that built the labor forces characterized by rational-

---

[15] Suchman (1995), p. 571.

[16] see Scott (1987), pp. 31–50.

[17] cf. Briggs (1968); Drucker (1954).

[18] see Martz (2013), pp. 387.

[19] Gomez (1981); Gomez and Probst (1999); Ulrich (1968); Ulrich and Probst (1988).

[20] Rüegg-Stürm and Grand (2019), p. 28.

[21] Katz and Kahn (1966); Yuchtman and Seashore (1967).

[22] Katz and Kahn (1966), p. 20.; Martz (2013), p. 388.

[23] Selznick (1948), p. 29.

[24] Selznick (1948)

[25] Scott (2003); Scott and Davis (2007); see Bilolo (2018).

ized and strictly controlled work.[26] This rational concept of organizations asked for control structures that could guarantee the intended high levels of efficiency, not only controlling for efficient transformation processes in production but also for a systematization of work to constrain social and labor unrest to guarantee economic and social order.[27]

From today's theoretical perspective, the question arises of how this systematization of organizational processes within the defined corporate boundaries could flourish. Where does legitimacy play its role in this rational view of organizations?

According to Shenhav (1995), in addition to increasing labor unrest that supported a shift towards new organizational structures and ideologies, it was the focus on establishing internal legitimacy by building up professional expertise and by concentrating on the progress of the organization that created the acceptance for this structural, controlled and hierarchical form of industrial organizations.[28] The "new hegemony of the engineer"[29] describes this internal view on organizations. Willing to function without external control by arguing in favor of the progress and efficiency gains that were reached and in contrast to current views on legitimacy, formalization, and hierarchy created a soil for generalized perceptions and assumptions that allowed the institutionalization of the rational view as "a legitimate canonical discourse during the Progressive Period (1900–1917)"[30].

The mentioned Progressive Period enabled the understanding and realization of the advancement of organizational systematization, which not only led to more standards and structures in production but also showed first spillover effects on social and other organizational issues, such as an engineering-focused social ideology.[31] Besides progressivism, the development created a nascent awareness of equality among employees, in which systems formed the guidelines of individual relationships in an organization, creating mutual responsibility and accountability and a "safeguard for the morality of organizations"[32] against unethical behavior.

In sum, the roots for understanding organizational legitimacy as a critical component for corporate survival originated from sociological research. However, the shift to consider organizations as open systems under the influence and observation from players located outside corporate boundaries was part of a more extended devel-

[26] see Drucker (1954); Taylor (1911); Thompson (1967).

[27] Shenhav (1995), p. 557.

[28] Shenhav (1995), p. 558.

[29] Haber (1964), p. 43, as cited in Shenhav (1995), p. 558.

[30] Shenhav (1995), p. 557.

[31] Shenhav (1995), p. 562.

[32] Kloppenburg (1986), as cited in Shenhav (1995), p. 564.

opment. Between the open system perspective and the described rational view, the natural view took its stake in explaining organizational structures and performance. According to Martz (2013), none of the described views can replace the other but deliver a guide to understanding and analyzing organizations.[33]

In his elaborations on legitimacy and bureaucracy, the famous sociologist Max Weber (1922) proclaimed that the legitimate order is a matter of authority whose structure influences the stability of social change. With his statements about validity that referred to the obedience to legitimate orders, though written with reference to a political context, Weber (1922/1978) already defined a sense of legitimacy that considers the social environment as an influential criterion.[34] In addition, with Weber (1922/1978), we get a first sense of the power of legitimacy when he described that orders are valid if they enjoy "the prestige of being considered binding, or, as it may be expressed, of 'legitimacy' "[35].

Weber concentrated on the formal instead of a behavioral approach to analyze organizations and critiques, potentially due to the different historical context at that time. Weber did not differentiate between rational and legal forms of legitimacy.[36] From today's standpoint, what is lacking in Weber's definition of legitimacy is its consideration of participation or, as Coleman (1990) stated,

> "away from elitist forms of legitimacy seems to have important effects on the members of society as persons [...] charged with shaping their own lives, they tend to become freer within themselves and more responsible and understanding in their regard for others and for society as whole"[37].

Though Weber can not be characterized as a rational, closed system theorist,[38] parallels exist in the sense that the rationalist view has limitations due to excluding human behavior.[39] The approach is further under scrutiny if internal and external conditions change and the variability of the environment asks for greater flexibility.[40] The naturalist view brings in this aspect of changing conditions by considering organizational adaptations becoming necessary to survive.[41] In addition, the human

[33] Martz (2013), pp. 386.

[34] Smith (1970), pp. 20–22; Weber (1922).

[35] Weber (1978), p. 31., as cited in see Bilolo (2018), p. 19.; Weber (1922).

[36] Smith (1970), p. 25.

[37] Coleman (1990).

[38] Scott and Davis (2007), as cited in Bilolo (2018).

[39] cf. Bennis (1959).

[40] see Coleman (1990), p. 29.

[41] see Bilolo (2018), p. 22.

element is further included and considers factors such as motivation and group dynamics as key components of organizational performance.[42]

However, what models exist to give the organizational environment a stake in evaluating organizational legitimacy? The following section describes the development towards open systems from the first theoretical narratives that started to consider the organizational environment to stakeholder and public value theory that give external stakeholders and the public a say in what, according to Suchman (1995), are "desirable, proper, or appropriate"[43] organizational actions.

## 3.2 The Ecological Level of Organizational Legitimacy—Considering the Organizational Environment With an Open System Perspective

The mentioned economic and social shifts drove theorists from a rational, closed system view to a natural system perspective that set the seed for an open system that takes the external environment and the public into account. A steadily evolving number of researchers focused on the influence of the environment on organizational actions and performance.[44]

Nevertheless, what is the environment of an organization by definition? Understanding the organizational environment allows us to dive deeper into its meaning for organizational research and legitimacy. For Scott (2007), the environment can be understood "as a store of resources as well as a source of opportunities and constraints, demands and threats"[45].

Some authors concentrated on the structural forces in the way that the environment can "create boundary spanning exigencies"[46] by an exchange of interdependencies with organizations.[47] Another strain of research puts social factors and the organizational environment into focus by stating that "organizations structurally reflect socially constructed reality"[48]. Authors such as Parsons (1956) were the first

[42] see Bion (1961); Mayo (1949).

[43] Suchman (1995), p. 574.

[44] see Aiken and Hage (1968); Berger and Luckmann (1966); Meyer and Rowan (1977); Parsons (1956); Thompson (1967).

[45] Scott and Davis (2007), p. 20.

[46] Meyer and Rowan (1977), p. 346.

[47] see Aiken and Hage (1968); Thompson (1967).

[48] Berger and Luckmann (1966), as cited in Meyer and Rowan (1977), p. 346.

to include the organizational environment and its value system as an influencing factor of organizational survival into theory.[49]

Similar to the rational closed system view, Parsons (1956) acknowledged the goal orientation of organizations and emphasized an efficient input-output transformation as one of its main tasks. However, he also saw organizations as "subsystems of a larger social system"[50] that reflect the subsystem's environment, in most cases representing society. In addition, he also defined the underlying value patterns of the organizational environment, the larger system, as a guiding principle for the orientation of organizations and characterized societal values as the "economic rational"[51] of organizations since profit-making is a success factor but not a functional criterion of society as part of a system.

How do these reflections by Parsons (1956) suit the developments of organizational legitimacy and further research that paved the way for theories such as stakeholder theory and public value theory that build the theoretical foundations of this dissertation?

Following Parsons (1956), legitimacy is created by the value patterns that are part of society and by the congruence of organizational goals that consider organizations as goal-directed systems, in which "every subgoal within the society must to some degree be integrated with the goal-structure of the society as a whole"[52]. This creates a guiding structure for organizations to receive legitimacy through a continuous assessment of organizational activities concerning their appropriateness and usefulness for societal goals.[53]

Consequently, Parsons (1956) laid the foundations for considering the environment as a guide for organizational actions by allowing the value system of the environment to create a normative structure without a binding legal form, going beyond corporate boundaries and becoming a frame of reference. This relates to the open system model that perceives organizations as social entities that are complex, dynamic, and embedded in and not separated from their environment. This environmental embeddedness leads to an active exchange between the organization and its environment. Organizations can be shaped and infiltrated by their environment, and some of the external elements can be at least as important as internal ones for the functioning of an organization.[54]

---

[49] Parsons (1956).

[50] Parsons (1956), p. 66.

[51] Parsons (1956), p. 68.

[52] Parsons (1956), p. 73, as cited in Bilolo (2018), p. 29.

[53] Bilolo (2018), p. 29 as in Pfeffer and Salancik (1978), p. 24.

[54] Scott and Davis (2007), p. 31.

Viewing organizations as systems engaged in interdependent activities with their environment makes its analysis quite complex. Scott (2007) called these coherences a "cultural soup"[55], which requires a concentration on the level of analysis in focus. In line with Parsons (1956) and Blau (1957), the analytical focus and the levels of analysis are crucial for characterizing organizational research.

For Blau (1957), the first level, the social-psychological level, shifts the lens on individual behavior within organizations. The second level covers the organizational level and focuses on the hierarchical structures of organizations, such as their working units and routines. This third level of analysis, the ecological level, emerged in the 1960s, taking the system of relationships and the organizations embedded in its environment into the center of analysis.[56]

For Parsons (1956), the organizational focus that should be considered is, first, the exchange processes of an organization that are mainly related to organizational resources and situational adaptation. The second focus covers the decision-making process of organizations and their internal goal-setting. The third focus relates to the ecological level of Blau (1957), which considers the alignment of organizational patterns with the "integrative exigencies of the society as a whole"[57].

This is why, since the 1970s, researchers have concentrated on the functioning of organizations under the influence of environmental effects. In their book on organizational ecology, Hannan and Freeman (1989) explored organizational success and selection in relation to its social structure. The authors highlighted that in the 1970s and 1980s, the most prominent American companies could not counteract foreign competition due to their ignorance of changing economic and social conditions. They further outlined that at that time, organizations acted as "passive brakes on social changes initiated by interest groups in the society"[58]. As a logical consequence, the authors emphasized that "changes in social, economic, and political systems do affect organizational structures and practices"[59].

This line of argumentation conforms with statements from authors, such as Meyer and Rowan (1977), who stated that organizational success and survival "depends on the ability of given organizations to conform to and become legitimated by environmental institutions"[60]. According to the authors, "organizations that incorporate societally legitimated rationalized elements in their formal structures maximize their

[55] Scott and Davis (2007).

[56] Blau (1957); see Scott and Davis (2007), pp. 17–18.

[57] Stern and Barley (1996), p. 151.

[58] Hannan and Freeman (1989), p. 5.

[59] Hannan and Freeman (1989), p. 6.

[60] Meyer and Rowan (1977), p. 352.

legitimacy and increase their resources and survival capabilities"[61]. However, what defines the rational elements that are socially legitimated, and how might these elements change in a dynamic environment? To answer this question, the following section dives deeper into the characteristics of the organizational environment with the help of the St. Gallen Management Model.

## 3.3 The St. Gallen Management Model as an Open System Model

Organizational research mentioning the limitations of the system model often referred to a disregard for external stakeholders and a clearer definition of the environment. Though included in the analysis, Martz (2013) criticized that stakeholders are not considered as elements contributing to organizational performance.[62] This requires bringing stakeholders as key elements of the organizational environment into the theoretical reflections of this dissertation.

The SGMM, as in Rüegg-Stürm (2019), postulates a systems-oriented view and that organizations rely on their embeddedness in the broader context that evolved over time.[63] Understanding organizations in their broader context as complex systems is a vital subject of the model. Interdependencies of elements exist whose interplay requires a careful understanding of its "prerequisites, capabilities, and dynamics"[64]. The SGMM not only takes the environment in a system-oriented view on organizations into its analysis but also defines the environment by relating to the environmental spheres and its involved stakeholders.

As the SGMM defines, environmental spheres are a point of reference when it comes to organizational value creation, and, as such, environmental spheres have to "continuously be analyzed to identify important changes and trends"[65].

As Fig. 3.1 shows, the economy, technology, nature, and society are building the environmental spheres of organizations. Part of these spheres that revolve around an organization are stakeholders that, as in the broadly accepted definition by Freeman (1984), are "any group or individual who can affect or is affected by the achievement of the organization's objectives"[66]. In the SGMM, these stakeholders have certain

---

[61] Meyer and Rowan (1977), p. 352.

[62] Martz (2013), p. 389.

[63] Pettigrew (1987) as in Rüegg-Stürm and Grand (2019)

[64] Rüegg-Stürm and Grand (2019), p. 30.

[65] Rüegg-Stürm and Grand (2019), p. 32.

[66] Freeman (1984), p. 46.

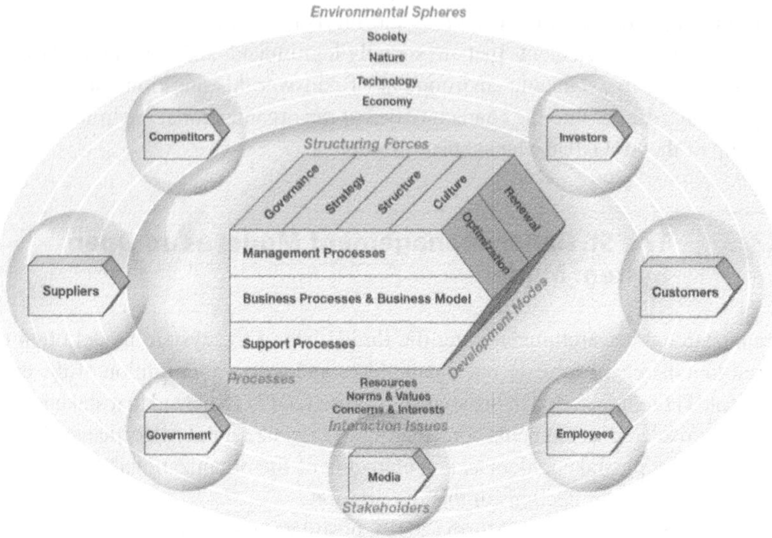


**Fig. 3.1** Environmental Spheres and its Stakeholders According to the SGMM (Rüegg-Stürm 2019)

expectations towards an organization and "confront an organization with different concerns, needs, and interests"[67].

Not only an organization itself is a complex system, but its environment with its dynamics requires organizational development and adaptation. Different factors contribute to environmental dynamics. Besides business-related factors such as innovation and creativity, political events, social dynamics, and crises nurture the organizational environment dynamics that require organizations to act and react. "Everyday organizational flux is optimized within the framework of these provisions"[68] that organizations take due to its influence on the fit of embeddedness into the organizational environment and, as later discussed, on relevant value criteria of organizations.

As mentioned, an environmental sphere comprises stakeholders with certain organizational expectations and concerns. Relative to its environmental sphere, these concerns and interests, due to the environment's dynamic, can be characterized as

[67] Rüegg-Stürm and Grand (2019), p. 32.

[68] Rüegg-Stürm and Grand (2019), p. 32.

fluid, be it a "homogeneous, heterogeneous, or even contradictory"[69] concern or interest. Before stakeholder concepts and their characteristics are described in the next chapter, the following sections illustrate what let those concerns, interests, and expectations become critical factors for organizational legitimacy and how they become part of organizational reality.

## 3.4 Organizational Legitimacy Resulting From the Sociology of Knowledge

Research on organizational legitimacy that considers the organizational environment raises the question of how research grasps what reality is and, in more detail, how concerns and interests of the environment can become an influential factor in organizational legitimacy in reality. It is vital to dive deeper into the relationship between reality, its related knowledge, and the phenomena related to the establishment of knowledge to answer how legitimacy is grounded in society.

According to Berger and Luckmann (1966), reality is defined as a "quality appertaining to phenomena that we recognize as having a being independent of our volition (we cannot 'wish them away')"[70]. The authors further emphasized the need for a "sociology of knowledge"[71] to be able to decide about the reality of knowledge being interpreted as "phenomena that possess special characteristics"[72]. In studying organizational legitimacy, answering questions on what processes lead to the establishment of knowledge in society and whether and how this knowledge is accepted is crucial. Though the definition of knowledge, its related ideas, and *Weltanschauungen* originate from a limited number of people, according to the authors, "everyone in society participates in its knowledge in one way or another"[73].

---

[69] Rüegg-Stürm and Grand (2019), p. 39.

[70] Berger and Luckmann (1966), p. 11.

[71] Berger and Luckmann (1966), p. 15.

[72] Berger and Luckmann (1966), p. 13.

[73] Berger and Luckmann (1966), p. 26.

Famous thinkers and researchers, such as Marx (1865),[74] Durkheim (1895/1982),[75] Nietzsche (1887/1998),[76] Weber (1922),[77] Scheler (1926),[78] Parsons (1956),[79] and Merton (1968)[80] shaped the understanding of knowledge and its sociology. Knowledge thereby does not relate to abstract theories but more to phenomena that are common sense[81] and that are taken for granted by society. Berger and Luckmann (1966) stated that it is this knowledge that "constitutes the fabric of meanings without no society could exist"[82].

The authors defined two stages of patterns becoming part of a society and its objective reality. Part of the first stage, institutionalization, is the habitualization of actions. Repetitive actions become a habit, reducing decision-making efforts due to known patterns that do not require an assessment from scratch. It leads to an institutionalization of habits that can be assigned to a particular type of action and its actors and, therefore, become part of society or a specific social group.[83]

Two main characteristics are essential for institutionalizing actions: its historicity and control system. As the first relates to a shared history that the mentioned typifications of repetitive actions create, the latter summarizes certain "predefined patterns of conduct"[84] that channel actions and their assessment with the help of a framework that is not legally formalized or part of a sanction mechanism. These control mechanisms exist in the public space called society. In sum, by linking the stage of institutionalization with the concepts of knowledge and reality, the typification of actions with the help of habitualization becomes knowledge. These institutionalized processes are considered reality as a matter of time and historicity.[85]

Different and specialized roles in society exist, and due to these different perspectives of its individuals and experts, it becomes difficult to establish "a stable symbolic

---

[74] Marx (1865).

[75] Durkheim (1895).

[76] Nietzsche (1887).

[77] Weber (1922).

[78] Scheler (1926).

[79] Parsons (1956).

[80] Merton (1968).

[81] According to Schuetz, socio-cultural *Lebenswelten* lead to the construction of common sense as part of the social approval of knowledge that is taken for granted by the majority of members of society (see Schuetz (1962)).

[82] Berger and Luckmann (1966), p. 27.

[83] Berger and Luckmann (1966).

[84] Berger and Luckmann (1966), p. 72.

[85] Berger and Luckmann (1966).

canopy for the entire society"[86] when it comes to the meaning of knowledge and its transition into objective reality. It is, therefore, the question, what knowledge is not bound to specific roles in society or its origins, and when can such knowledge become an accepted reality?

Berger and Luckmann (1966) spoke of a reification process that describes the process towards objective reality when it comes to the "apprehension of products of human activity as if they were something other than human products—such as facts of nature, results of cosmic laws, or manifestations of divine will"[87]. Consequently, knowledge "is raised to the level of a relatively autonomous sub-universe of meaning"[88], and "can act upon the collectivity that has produced it"[89].

After exploring the institutionalization of knowledge, legitimation is the second step that leads to developing an objective social reality, which brings us back to the center of organizational legitimacy. Legitimation "is this process of explaining and justifying"[90]. As a cognitive process, the objectivated meaning of social phenomenons is justified as valid by, as Berger and Luckmann (1966) stated, "giving a normative dignity to its practical imperatives"[91]. This second step combines a cognitive process with its normative character. It relates values to knowledge since the Why behind organizational actions is linked with the facts the social world has created by its institutionalization.

Explaining the institutionalization and legitimation of knowledge becoming objective reality might sound like a smooth and clearly defined societal process. However, it must be conceded that objective reality is always under scrutiny. As Berger and Luckmann (1966) assessed, "keeping chaos at bay, all social reality is chaos, all societies are constructions in the face of chaos"[92]. Consequently, how is it possible to delineate the space in which objective reality is created? Where does the institutionalization and legitimation of knowledge occur, or in other words, what is society?

[86] Berger and Luckmann (1966), p. 103.
[87] Berger and Luckmann (1966), p. 106.
[88] Berger and Luckmann (1966), p. 103.
[89] Berger and Luckmann (1966), p. 103.
[90] Berger and Luckmann (1966), p. 111.
[91] Berger and Luckmann (1966), p. 111.
[92] Berger and Luckmann (1966), p. 121.

## 3.5 Society and the Public as a Construct of Meaning and Communication

Learning from Berger and Luckmann (1966) about the role knowledge plays in becoming institutionalized and legitimated as objective reality among members of institutions that the authors call societies triggers the question of how society is defined and what makes a society a society. Luhmann (2012) emphasized that "society is a meaning-constituting system"[93], while "meaning is accordingly an infinite and hence indeterminable referential complex that can be made accessible and reproduced in a determined manner"[94]. Definitions and thoughts about the concept of society and social life range back to Plato's Republic and Laws dating back to the 4th century B.C.E., to Confucius in the 5th century B.C.E. with his fundamental theories on social relationships, and to other famous thinkers. According to the dictionary Merriam-Webster, several definitions of the term society exist. A selection of quotations defines society as "a voluntary association of individuals for common ends"[95], "an enduring and cooperating social group whose members have developed organized patterns of relationships through interaction with one another"[96], "a part of a community that is a unit distinguishable by particular aims or standards of living or conduct"[97], "a social circle or a group of social circles having a clearly marked identity"[98], and a "community, nation, or broad grouping of people having common traditions, institutions, and collective activities and interests".[99]

Habermas' (1992/1996) definition of civil society revealed the link between society and the public sphere by stating that

> "civil society is composed of those more or less spontaneously emergent associations, organizations, and movements that, attuned to how societal problems resonate in the private life spheres, distill and transmit such reactions in amplified form to the public sphere. The core of civil society comprises a network of associations that institutionalizes problem-solving discourses on questions of general interest inside the framework of organized public spheres"[100].

---

93 Luhmann (2012), p. 21.

94 Luhmann (2012), p. 21.

95 Merriam-Webster (n.d.-a).

96 Merriam-Webster (n.d.-a).

97 Merriam-Webster (n.d.-a).

98 Merriam-Webster (n.d.-a).

99 Merriam-Webster (n.d.-a).

100 Habermas (1996), pp. 366–367.

This definition already focuses on the power of discourse and on the resonating of societal issues that rest on the private sphere and its individuals that is later discussed in chapters dealing with the individual basic needs and public value.

The public is, therefore, the social environment that is more than the sum of its members, and at the same time, a result that its members as human beings experience in an active process. The public builds up a "reference point for action"[101] and includes concepts such as society and community. It is an abstraction that "is generated on the basis of experiences made in daily practices, analytical insight, and all sorts of projections as to complex phenomena"[102].

The definitions of society and the public appeal to characteristics such as action and interaction to distinguishable patterns such as certain conduct, reference points, experiences, identity, activities, or interests. Nevertheless, how are the definitions of society related to knowledge and the creation of an objective reality?

One often referred publication is Tönnies (1887/2017) and his postulates about community and society and their relation to knowledge and the will of human beings. According to the author, "the modern subject desired above all a clear and certain knowledge of the effects of all possible actions"[103], the most basic form of knowledge includes knowing "what was valuable and how it might be obtained"[104]. The author defined "Gesellschaft" as a society and "Gemeinschaft" as a community. In contrast, both include social actions due to a moral will, whereas their members "share a common consciousness, an agreement over ideals"[105].

Luhmann (2012), while drawing from his system-oriented perspective, emphasized that society is a complex system. Luhmann challenges traditional definitions of society. Definitions of society that the author criticizes refer to characteristics such as a sort of consensus among its members "by concordant opinion and complementary purpose"[106] that constitutes society. He further questioned if, in a global world, societies have to be constrained to regional characteristics as many definitions include[107]. Luhmann (2012) crystallized his definition of society with a focus on the main structuring force of society: communication. For Luhmann (2012), society is

[101] Meynhardt (2009), p. 204.

[102] Meynhardt (2009), p. 204.

[103] Tönnies and Loomis (1887), p. xvii.

[104] Tönnies and Loomis (1887), p. xvii.

[105] Tönnies and Loomis (1887), p. xxii.

[106] Luhmann (2012), p. 6.

[107] Luhmann (2012), pp. 9–11.

communication while "everything impacting society from outside that is not communication must therefore have passed through the double filter of consciousness and the possibility of communication"[108].

Communication leads to specific labeling and to distinctions that, according to Bechmann and Stehr (2002) in their description of Luhmann's work, "are the start of all perception and recognition"[109]. With its habits and objective reality, the institutionalized world reveals a certain consistency in understanding a "logic" that relates to associations people have concerning organizations and their functions. This consistency that establishes the logic results from a reflective consciousness created with the help of communication and language.

Language and communication create a logic of objective reality with the help of meaning. In that case, it is the question of how communication can unfold its impact on society and organizations to create meaning. Language is a crucial factor "for the detachment from the 'here and now' of everyday life"[110] that knowledge can become common sense and objective reality. Further information and understanding drive the establishment of meaning in society. Luhmann (2012) highlighted that an ongoing flow of communication, which ends in acceptance or rejection, is part of the understanding process,[111] while understanding combines past inferences with current references.[112] The author emphasized that a precondition for the development of information is interest since "it cannot be gained purely passively as a logical consequence of signals from its environment"[113]. Interest and expectations in information consisting of informational value can change the state of its underlying system, which, in the organizational context, can have a massive impact on the understanding and interpretation of organizational actions and underlying expectations.

Consequently, while common sense and knowledge are not stable but fluid concepts, the question arises of how the knowledge that becomes part of the objective reality in society can challenge organizations and their legitimacy due to a shift of discourse.

---

[108] Luhmann (2012), p. 63.

[109] Bechmann and Stehr (2002), p. 69.

[110] Bechmann and Stehr (2002), p. 55.

[111] Luhmann (2012), p. 137.

[112] Luhmann (2012), p. 37.

[113] Luhmann (2012), p. 36.

## 3.6 Discursive Shifts in Society Challenging Organizational Legitimacy

Shifting the perspective closer to the organization and its environment raises the question of how a change of discourse in society through communication can become a matter of organizational legitimacy and lead to organizational change. Before diving into the definition of discourse, the characteristics of organizational change and its relationship with management responsibilities are first presented.[114]

Drucker (2001) answered to whom the management of organizations is accountable with the three responsibilities of organizations. These are, first, economic performance as a central social function and purpose for a business enterprise. The measure for the first responsibility lies in financial performance metrics that are easily accessible. The second responsibility describes the link to the individual by acknowledging that an organization's crucial resource is its people. Third, social impacts stand in the center since, according to Drucker (2001), "free enterprise cannot be justified as being good for business; it can justified only as being good for society"[115]. This links managerial responsibility with organizational legitimacy and its connection to society since organizations are not independent of the shifts and debates that dominate the public.

Imerman (2018) spoke of cycles of institutional legitimacy expressing specific pressures and a temporal dynamic of these pressures for organizations to adapt. However, there needs to be a systemic approach that accounts for the fluidity and cyclicity of legitimacy.[116] To acknowledge the dynamic characteristics of legitimacy and its effect on organizations, we first have to explore at what levels of analysis organizations are affected when organizational change is required.

The SGMM states that change triggers often relate to new opportunities that organizations seek but also to shifting expectations of society and stakeholders and to difficulties that exist in terms of organizational value creation.[117] It shows that the internal and external world can impact organizations.

Pettigrew (1987) developed a framework that relates to those different levels of analysis that are affected by or the cause of organizational change. According to the author, "the process of change refers to actions, reactions, and interactions from the various interested parties as they seek to move the firm from its present

---

[114] For additional research on organizational change, see Barnett and Carroll (2003); DiMaggio and Powell (1983); Weick and Quinn (2003).

[115] Drucker (2001), p. 965.

[116] Imerman (2018).

[117] Rüegg-Stürm and Grand (2019), p. 65.

to its future state"[118]. As Fig. 3.2 reveals, three dimensions impact organizational change. First, the content dimension describes what is behind organizational change, while second, the process dimension relates to how change is conducted. Third, the context dimension focuses on the Why with an inner and outer context. The inner context includes the organizational structure and culture, while the outer context refers to the economic, political, and social environment. The outer context can act as a legitimating force that either confirms the existing order or shapes a new one. It is especially crucial to observe organizational change in times of ideological shifts. Ideological shifts gathering momentum can result in organizational change.[119]

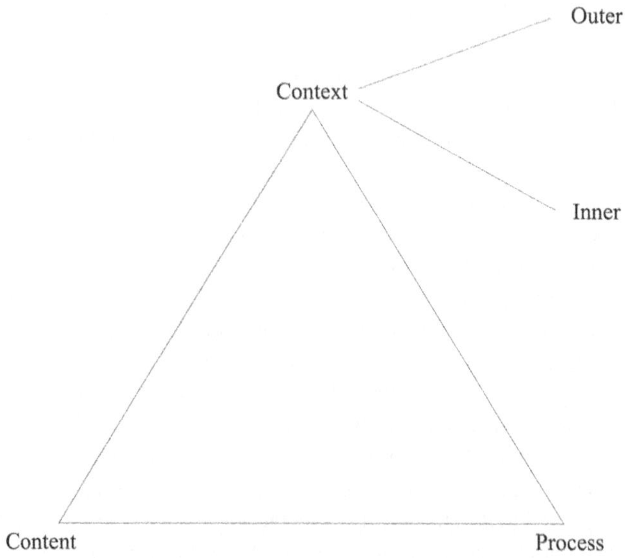


**Fig. 3.2** Organizational Change and its Levels of Analysis (Pettigrew 1987)

Part of the outer context are societal expectations leading to organizational change with the help of communication and related information. Under the concept of discourse and with the help of communicated information, the environment of organizations exerts its influence on organizations. It puts the concept of discourse into focus.

[118] Pettigrew (1987), p. 658.
[119] Pettigrew (1987), p. 664.

Discourse familiarizes unfamiliar practices, leading to understanding and acceptance[120]. Discursive action, therefore, is "the production and diffusion of texts that create or transform the cognitive schemas through which actors interpret and give meaning to reality"[121]. Discourse is a concept that is well-explored in research. Its prerequisites and conceptional footings refer to deliberation as "a process through which participants address their conflicts, share information, exchange arguments and make decisions"[122]. The arena of deliberation is the public sphere, as described in the previous sections.

Several authors and thinkers, such as Durkheim (1895),[123] Mead (1934),[124] and Habermas (1996)[125] contributed to the development of the concept of discourse and explored its influence in the public sphere. In his conceptual framework, Mead (1934) referred to normatively regulated communication as a mediator of interaction. He stated,

> "we are what we are through our relationship to others. Inevitably, then, our end must be a social end, both from the standpoint of its content... and also from the point of view of form. Sociality gives the universality of ethical judgments and lies back of the popular statement that the voice of all is the universal voice; that is, everyone who can rationally appreciate the situation agrees"[126].

As stated in Habermas (1987), Mead made its projection of a "communicatively rationalized society"[127], while Durkheim referred to a collective consciousness and reflections on morality. Morality for Durkheim does not only include obligatory characteristics but also desirable ones. In his reference to Durkheim, Habermas (1987) emphasized "the morally good is at the same time something worth pursuing; it could not be effective as an ideal and arouse enthusiastic fervor if it did not offer the prospect of satisfying real needs and desires"[128]. In this statement, we see the link to the needs of individuals that will later lead to public value theory being further explored in Sect. 5.2.

---

[120] Lounsbury and Glynn (2001), as cited in Debenedetti et al. (2021), p. 334.

[121] Phillips et al. (2004), as cited in Debenedetti et al. (2021), p. 334.

[122] Palazzo and Scherer (2006), p. 80.

[123] Durkheim (1895).

[124] Mead (1934).

[125] Habermas (1996).

[126] Mead (1934), as cited in Habermas (1987), p. 94.

[127] Habermas (1987), p. 94.

[128] Habermas (1987), pp. 48–49.

A discourse ethics was developed by Mead (1934), describing a universe of discourse "which transcends the specific order within which the members of the community, in a specific conflict, place themselves outside of the community order as it exists, and agree upon changed habits of action and a restatement of values"[129]. The universe of discourse, with its rational procedures, describes a thought process on which action follows. Mead (1934) further contributed that

> "rational procedure, therefore, sets up an order within which thought operates; that abstracts in varying degrees from the actual structure of society … It is a social order that eludes any rational being who is or may be in any way implicated in the situation with which thought deals. It sets up an ideal world, not of substantive things, but of proper method. Its claim is that all the conditions of conduct and all the values which are involved in the conflict must be taken into account in abstraction from the fixed forms of habits and goods which have clashed with each other. It is evident that a man cannot act as a rational member of society, except as he constitutes himself as a member of this wider common world of rational beings"[130].

In an organizational context, this statement shows how a widely accepted conduct can develop out of discourse that includes values that are part of a rational procedure acting as a reference for organizations. Part of the essence of societal discourse are different constructs that can shape the direction of organizational legitimacy. Changing stories, paradigms, or narratives can affect mindsets, purposes, paradigms, or perspectives.[131] In this context, narratives relate to "preferences, expectations, identities and definitions of situations [that] are seen as arising from interactions within a social system, thus as embedded in social norms and cultural conventions of discourse"[132]. Mindsets and perspectives are based on values and knowledge that shape the world and influencing actions.[133]

A final question remains that elaborates on what builds the relationship between the concept of discourse and organizational legitimacy. Legitimacy is a result of the deliberative process in the public sphere. Competing discourses shall lead to collective outcomes due to a willingness to strive for mutual understanding.[134] Legitimacy "is based on the institutional design of discursive arenas and the procedural design

---

[129] Mead (1934), as cited in Habermas (1987), p. 92.

[130] Mead (1934), as cited in Habermas (1987), p. 92.

[131] see Schot and Steinmueller (2018).

[132] March (1996), p. 285.

[133] see Gorddard et al. (2016); Moore et al. (2018); Waddock (2020).

[134] see Dryzek (2001), Habermas (1996).

of public will-formation"[135]. Consequently, most concepts of organizational legitimacy refer to the goal that organizational actions become stable and comprehensible and that organizations gain support from their social environment by its normative conformity with the external context.[136] Related to the different forms of legitimacy as presented by Suchman (1995), organizations have to build up their legitimacy "to make its products cognitively understandable, morally acceptable and legally compliant"[137].

What sounds trivial represents a current challenge for corporate strategies, actions and their disclosure. Existing concepts of organizational legitimacy are challenged by current societal developments, its discourse, and the influence of public will formation. The strategic approach, for example, as defined by Suchman (1995), includes "a high level of managerial control over the legitimation process"[138]. It contains a direct manipulability of legitimacy that, as Suchman (1995) already acknowledged, becomes more difficult to ensure when moving from the pragmatic form of legitimacy to the moral and cognitive form.[139] The cognitive form, however, relates to legitimacy within a geographical and cultural context. Palazzo (2006) emphasized that the cognitive form is nowadays challenged by the development of a pluralist modern society that "erodes the normative taken-for-grantedness"[140]. Consequently, the author spoke in favor of a greater focus on the moral form of legitimacy and factors that relate to the outer context of organizations that directly take the construct of societal discourse and its impact on organizational legitimacy into account. Palazzo (2006) stated that "moral legitimacy becomes the decisive source of societal acceptance for corporations in an increasing number of situations"[141].

This speaks for a perspective shift that puts greater emphasis on the societal discourse since pluralism and deliberation is giving further prominence to the public discourse of civil society that do not allow a taken-for-grantedness of societal rules, but require the incorporation of social issues and values with that are represented by the organizational environment and its stakeholders. Being open for this deliberative approach to gain legitimacy to be able to sustain a business means accepting

---

[135] Dryzek (1999), p. 43; Habermas (1996), as cited in Palazzo and Scherer (2006), p. 81.

[136] see Carroll (1979); Palazzo and Scherer (2006); Strand (1983); Suchman (1995).

[137] Debenedetti et al. (2021), p. 333.

[138] Suchman (1995), p. 576.

[139] Suchman (1995).

[140] Palazzo and Scherer (2006), p. 74.

[141] Palazzo and Scherer (2006).

an increasing number of legitimacy challenges stemming from the moral form of legitimacy by anticipating and actively managing discursive shifts.[142]

Before exploring the environment of organizations, the theoretical background of its stakeholders as the main influencers of societal discourse, the question of the cause of temporality and why we speak about shifts of discourse is part of the following section.

## 3.7 The Temporal Aspects of Discursive Shifts and its Related Issues

As acknowledged in the previous sections, legitimacy is characterized by its circularity and fluidity.[143] Imerman (2018.) stated that "legitimacy is an inherently dynamic, relational concept that is sensitive to existing social and material conditions, which are constantly evolving"[144].

Berger and Luckmann (1966) already elaborated on this temporality of meaning and systems that are stable and, at the same time, dynamic, being replaced by new systems frequently, which requires distinction since "the temporal contingencies of all attitudes towards the world are now much more strongly assumed"[145]. One special challenge the authors illustrated is the transmission of knowledge from older to younger generations. A successor generation can challenge the legitimacy of an objective reality. Preceding generations understand the world as they have shaped it, but "the salient elements of the institutional tradition"[146] needs explanation and justification that the objective reality will be accepted by succeeding generations that "the objectivity of the institutionalized world 'thickens' and 'hardens'"[147]. The temporal aspects of organizational legitimacy challenge organizations to anticipate the future and, simultaneously, "carefully and attentively approaching and appreciatively considering the past"[148].

Berger and Luckmann postulated (1966) that the process towards objective reality is a question of whether human kind "retains the awareness that, however objectivated, the social world was made by men, and therefore, can be remade by

[142] Palazzo and Scherer (2006).

[143] Imerman (2018).

[144] Imerman (2018), p. 75.

[145] Berger and Luckmann (1966), p. 24.

[146] Berger and Luckmann (1966), p. 111.

[147] Berger1966, p. 111.

[148] Rüegg-Stürm and Grand (2019), p. 66.

[it]"[149]. The temporal aspect of organizational legitimacy and the transmission of objective reality to younger generations become especially intriguing when looking at the concept and development of sustainability. The grand challenges the world faces open the discussion of whether discursive change can shape a new objective reality that allows humankind to remake the world or if nature lets humans become a product of it, with whatever consequences it might have.

According to this debate, part of the discourse that Berger and Luckmann (1966) called salient elements are societal issues that organizations must consider to gain or maintain legitimacy. According to Hoffman and Ross (1999), "issues define what the field is, making links that may not have previously been present"[150]. For organizations, issues become salient if the topics related to the issues are institutionalized. They become a matter of organizational legitimacy if the issues reach the status of standards that are able to describe the behavior of a reference group[151]. In other words, issues become salient if the issues and positioning towards them turn into a social obligation.[152]

Organizations are strategically challenged since the environment and its stakeholders may have different views on an issue, directly leading to questioning organizational legitimacy. Consequently, the actors in discursive shifts play a crucial role in shaping organizational legitimacy. Social stakeholders shape society. Therefore, the theoretical footings of stakeholder theory are further elaborated in the following chapter and sections.

---

[149] Berger and Luckmann (1966), p. 111.

[150] Hoffman and Ross (1999), p. 352., as cited in Debenedetti et al. (2021).

[151] Philippe and Durand (2011), p. 970., as cited in Debenedetti et al. (2021).

[152] see Debenedetti et al. (2021); Hoffman and Ross (1999).

# 4 Stakeholder Theory—Shaping Organizational Legitimacy

Relating to the role that the outer context and environment of organizations play in establishing and maintaining organizational legitimacy, this chapter deals with the relevant players that are part of the environment. Organizations face the challenge of understanding the varying interests of stakeholders, the topics and issues that drive a corporate-stakeholder relationship, and its value relevance. Being confronted with a range of societal issues while increasing societal awareness is developing, stakeholder relationships and dialogue "legitimize a 'corporate license to operate' "[1].

In contrast to the past and the strategic legitimacy approach, organizations are less able to trigger some symbolic actions to satisfy stakeholder expectations. In contrast, understanding the underlying issues and possible reasons for conflict with the parties who can contest organizational legitimacy turns into a critical corporate competence. These circumstances include organizational actions and attitudes that incorporate, as Jonker and Nijhof (2006) stated, a "looking through the eyes of others"[2] to be able to evaluate and manage expectations.

However, different views exist concerning the perception of this organizational stakeholder relationship.[3] Exploring the relationship further with a view on discourse ethics and deliberation, as discussed in the previous chapter, means that organizations and their stakeholders are part of a relationship with a "forceless force of the better argument"[4] that includes communication and argumentation for creating an understanding and an outcome. Communication also focuses on a postmodern

[1] Heath et al. (2006), as cited in Høvring et al. (2018), p. 627.

[2] Jonker and Nijhof (2006), p. 456.

[3] Apart from discourse ethics and the critical hermeneutic perspective and a postmodern perspective, the liberal humanist perspective focuses on the concept of shared values, shared meaning, and consensus (see Isaacs (1993)).

[4] Habermas (1998), as cited in Høvring et al. (2018), p. 630.

 
T. Heroth, *Corporate Sustainability Disclosure Matters*,
https://doi.org/10.1007/978-3-658-47653-3_4

perspective of stakeholder dialogue that concentrates on a new understanding of developments in the world, such as social change. As mentioned before by Luhmann (2012), it is not the consensus that is a precondition for building a society but more the otherness and openness for new perspectives with the means of communication. Therefore, conflict, confrontation, and not taken-for-grantedness are likely part of organizational stakeholder interactions that evolve around the meaning of business and organizational legitimacy.[5]

The concept and meaning of stakeholders are also contested. As presented in the following section, several theoretical approaches emphasize opposing views on the importance and the power of shaping organizational legitimacy.

## 4.1 Stakeholder Theory and its Contesters

Stakeholder theory is a vital research strain and a theoretical foundation of strategic management research. As mentioned, a stakeholder is "any group or individual who can affect or is affected by the achievement of the organization's objectives"[6]. Companies are part of a network of coalition partners whose interests determine the success of a company's products and services. Research in strategic management focuses on stakeholders, such as "investors/stockholders, employees, managers, customers, and the community (that) may have direct impact on the successful implementation of a corporate entrepreneurship strategy"[7].

Freeman (1984, 1994) was one of the first who defined the interaction between companies and their stakeholders and the dynamics behind this relationship. The goal was to define a normative theory of why stakeholders matter and to develop ethical principles for corporate stakeholder interaction. This normative approach referred to the moral obligations towards stakeholders.[8]

Other approaches towards stakeholder theory that challenge the normative view include an instrumental and descriptional perspective.[9] The descriptive aspect refers to "specific corporate characteristics and behaviors"[10] that support active stakeholder management of companies. The instrumental aspect considers the goal and

[5] see Burchell et al. (2013); Deetz and Simpson (2004); Glozer et al. (2013); Høvring et al. (2018).

[6] Freeman (1984), p. 46.

[7] Kuratko et al. (2007), p. 57.

[8] see Freeman (1984); Freeman (1994), Mitchell et al. (1997).

[9] cf. Donaldson and Preston (1995).

[10] Donaldson and Preston (1995), p. 70.

practical-oriented view on stakeholder management to achieve "traditional corporate objectives"[11], such as financial performance. It further highlights that stakeholder management and mutual trust and cooperation will lead to competitive advantage and improvements in the corporate bottom line.[12]

Descriptive studies in the 1970s and 1980s already showed results of surveys that support an existing stakeholder sensitivity of managers.[13] In the 1980s and 1990s, national regulation and legal developments put stakeholders into focus, such as the impact of corporate takeovers on players other than shareholders in the US and the implementation of employee representation in boards of directors in Germany.[14]

Most studies that are related to stakeholder theory referred to this "tripartite typology"[15] of normative, descriptive, and instrumental views.[16] However, Donaldson and Preston (1995) themselves questioned the descriptive and instrumental view in a way that they perceived stakeholder theory as normative since only the moral obligations of managers towards their stakeholders are the only real argument and valid basis for the existence of a stakeholder theory.[17]

Freeman (1999) concluded that we should better drop the tripartition since "the very idea of a purely descriptive, value-free, or value-neutral stakeholder theory is a contradiction in terms"[18]. Freeman et al. (2020) suggested that stakeholder theory "reflects a coherent worldview of business that reflects a broader awareness about its role, purpose, and full range of immediate and long-term impacts on society"[19].

Still, various opposing views exist that still consider shareholder value as the primacy[20] and focal point of a company and perceive stakeholder management under the notion of stakeholderism as an illusionary promise that only fills a managerial agenda without any benefits.[21] The development of the shareholder value paradigm

[11] Donaldson and Preston (1995), p. 71.

[12] see Berman et al. (1999); Jones (1995).

[13] see Brenner and Molander (1977); Donaldson and Preston (1995); Posner and Schmidt (1984).

[14] see Donaldson and Preston (1995).

[15] Freeman (1999), p. 234.

[16] see Butterfield et al. (2004); Rose et al. (2018); Theodoulidis et al. (2017); Vitolla et al. (2019a); Vitolla et al. (2019b).

[17] Donaldson and Preston (1995), pp. 87–88.

[18] Freeman (1999), p. 234.

[19] Freeman et al. (2020), p. 219.

[20] see Blair and Stout (1999), p. 253.

[21] cf. Bebchuk and Tallarita (2020), Bebchuk et al. (2020).

as the single "corporate objective function"[22], seen in its historical context, developed out of focus on organizational growth, the expansion into new markets, and the hostile takeover movement in the US and the United Kingdom in the 1970s and 1980s.[23] These hostile takeovers empowered shareholders who gained a greater say and lever to control, appraise, or punish management for underperformance. It led to metrics such as earnings per share (EPS) or the return on capital employed (ROCE) that, according to Deakin (2005), allowed a benchmarking of corporate performance out of the shareholder perspective.[24]

As a positive consequence, Rappaport (1981), in his contribution to shareholder value, triggered an improvement of financial disclosure regulations and a simplification and practical-oriented modification of corporate valuation.[25] Consequently, as the name tells, the concept of shareholder value puts its singular focus on the shareholder and investor as the dominant stakeholders of a company and perceives the notion of value as a monetary metric without taking the social and moral values of businesses and related individuals into account.[26]

Challenging stakeholder theory is trivial because stakeholder concepts are contested.[27] Theoretically, those concepts take time to grasp their interpretability. Its contestability is due to a need for empirical evidence that can be reduced by a more detailed view of corporate stakeholder relationships and their value relevance in practice.[28]

Empirical evidence can be gathered by a detailed look at the organizational environment and its stakeholders from a normative perspective. However, how are these stakeholders so far classified and identified in research? The following section includes the classification of stakeholders from a managerial perspective.

## 4.2 Managerial Perspectives and the Classification of Stakeholders

Stakeholder theory most often relates to attributes and determinants of corporate stakeholder relations from a corporate and management perspective. Mitchell et al. (1997) proposed a typology for identifying stakeholders based on managerial

[22] Deakin (2005), p. 11.

[23] see Deakin (2005); Drucker (2001).

[24] Deakin (2005), p. 14.

[25] see Bilolo (2018), p. 112.; Rappaport (1981).

[26] see Bilolo (2018). p. 113.

[27] cf. Bonnafous-Boucher and Rendtorff (2016); Freeman et al. (2020); Miles (2017).

[28] see Miles (2017).

perception and three different attributes: power, legitimacy, and urgency. The legitimacy and time-related focus suit the abovementioned characteristics of organizational legitimacy as presented.[29]

According to the authors, legitimacy analyzes the existence of specific claims postulated by stakeholders. Power describes the ability of a stakeholder to influence a company. Urgency puts the before-mentioned dimensions of power and legitimacy in a time-sensitive or dynamic perspective. It adds a claim-related appraisal of the criticality level of an issue raised by stakeholders.

The theoretical framework by Mitchell et al. (1997) should enable the exploration of the salience of stakeholders. Salience in this context can be described as "the degree to which managers give priority to competing stakeholder claims"[30] and refers to the attention managers pay to specific entities of the corporate environment. Various researchers based their research on the framework by Mitchell et al. (1997) and the managerial perspective on stakeholder relationships.[31] The concept brought the corporate-stakeholder relationships and issues into focus.

As Miles (2017) summarized in her stakeholder theory classification of 885 stakeholder theory definitions, managerial perceived determinants focus on the identification and relevance of stakeholders as defined by companies. Subject for deriving a stakeholder concept that focuses on identifying stakeholders by companies in the form of interest or claim that stakeholders raise, the type of relationship, the basis of legitimacy or its risks and power attributes. [32]

Criteria that are more focused on the outcome of a corporate-stakeholder relationship cover the strategic focus, the form of interaction, and factors such as frequency of contact and proximity of a relationship.[33] Such research delivered crucial knowledge about corporate stakeholder relationships that also suit the research of this dissertation. Stakeholder orientation, in this context, draws from existing definitions by referring to the degree to which organizations incorporate stakeholders into corporate disclosure. According to Amiri et al. (2022), stakeholder orientation "boosts stakeholder management capacity by enabling closer interaction with stakeholders to integrate their interests into organizational decisions" [34]. Stakeholder orientation creates an understanding of stakeholders' interests, influence, and legitimacy

[29] Mitchell et al. (1997).

[30] Mitchell et al. (1997), p. 854.

[31] see Abdel-Maksoud et al. (2021); Buniamin (2020); Rodrigue et al. (2013).

[32] Miles (2017), p. 444.

[33] Miles (2017), p. 447.

[34] Amiri et al. (2022), p. 386.

claims.[35] It helps to answer the question of social stakeholders' role in terms of a certain value relevance of organizational legitimacy and is thereby grasped from a view of a bidirectional relationship between organizations and stakeholders, as Jones (2011) advocated.[36]

Consequently, there needs to be more research related to stakeholder theory that includes an empirical approach to explore the value relevance of organizational stakeholders due to organizational legitimacy and discursive shifts in society. How can organizations reach organizational legitimacy by social stakeholder orientation and a salient issue-related focus on public value? Although "the discursive approach to organizational legitimacy will have to deal with the growing importance of dissensus communication in pluralizing and globalizing societies"[37], how can organizations proactively anticipate and implement organizational actions that align with societal discourse and satisfy the needs of society and its stakeholders? The link to concepts and theories, such as public value theory and its empirical investigation, must be further explored in research. Many publications concentrate on the characteristics of a relationship instead of the issues that form and influence the corporate stakeholder relationship.

This dissertation focuses on social stakeholders and the societal issues that evolve in societal discourse that build a reference base for organizational legitimacy. Social stakeholders are those stakeholders that have a high exposure to the social pillar and its subtopics. These include the stakeholder society, community, employee, NGO, association, and media. A second group of stakeholders, financial stakeholders, is further considered to compare its value relevance with the group of social stakeholders and with the help of sustainable and financial performance metrics. Financial stakeholders are those in a financial relationship with a company based on investments or linked to credits. This stakeholder group contains the stakeholder investor, bank, and shareholder.

With a differentiation between two different categories of stakeholders, this dissertation does not aim to prioritize specific stakeholders. On the contrary, categorizing stakeholders is solely due to empirical reasons and to unpuzzle those stakeholders' value relevance, which recently increasingly attracted attention in discussions about corporate legitimacy as part of a discursive shift. It is more to bring social stakeholders in than to exclude financial stakeholders. Since discursive shifts increase the attention towards social stakeholders and related issues, the value relevance of

[35] see Bettinazzi and Zollo (2017); Bettinazzi and Feldman (2021); Harrison et al. (2010); Zollo et al. (2009).

[36] see Jones (2011); Jones et al. (2018).

[37] Palazzo and Scherer (2006), p. 73.; Steinmann and Scherer (1998).

specific stakeholders, such as society and community, might change. This requires empirical evidence and a more detailed view of societal discourse, its increasing value relevance, and an understanding of what organizational legitimacy means in the current organizational context, as explored in this dissertation.

Detecting the value relevance of stakeholder orientation and public value topic salience means retrieving a deeper understanding of the concept of value with a focus on public value theory, making organizational legitimacy a measurable criterion, as presented in the following chapter.

# 5 Public Value Theory—Measuring Organizational Legitimacy

Organizational legitimacy and stakeholder theory are closely linked with the concept of public value. Thus, public value as a concept not only integrates well into the theories and dynamics of organizations presented so far but also contributes to the detection of how stakeholders as part of the environment and the societal discourse are involved in and impact organizational legitimacy and its value relevance.

Though an abstract construct, the public becomes a reference point for corporate actions and can create value through this process.[1] With increasing societal expectations exerted on organizations, the value creation process is asked to be more closely aligned with the public. Public value can, therefore, be understood as "a shared or collectively held value about the quality of a relationship involving the public"[2].

Public value has its primary roots in the context of public policy and public management domain, which is not the subject of this dissertation.[3] Several additional domains were tackled by ongoing research on public value. These views focus on public value in domains such as economics[4] and philosophy.[5]

In line with the focus of this dissertation, public value, according to Benington and Moore (2011), "is now at the forefront of cross-national discussion about the changing roles of the public, private, and voluntary sectors in a period of profound political economic, ecological, and social change"[6]. Existing publications related

---

[1] Meynhardt (2015), p. 151.

[2] Meynhardt (2015), p. 151.

[3] For a study of public value and its origins and link to public policy, see Alford and Hughes (2008); Benington (2009); Benington and Moore (2010); Bozeman (2007); Meynhardt (2009); M. H. Moore 1995; Talbot (2009).

[4] Hutton et al. (2007).

[5] Morrell (2009).

[6] Benington and Moore (2011).



T. Heroth, *Corporate Sustainability Disclosure Matters*,
https://doi.org/10.1007/978-3-658-47653-3_5

to this research focused on measuring public value[7] and on accounting and corporate disclosure.[8] The development of sustainability and its growing importance as part of the corporate agenda links public value with the private sector and with new opportunities for researchers concerning its measurability, its connections with organizational legitimacy, and the role of stakeholders. However, measuring public value means knowing about the concept of value, its components, and its driving forces. Therefore, the following section dives into the concepts and definitions of value.

## 5.1 Existing Concepts and Definitions of Value

Various definitions exist concerning the concept of value. Many of these definitions relate to the price component by describing value as "the monetary worth of something"[9] or "a fair return or equivalent in goods, services, or money for something exchanged"[10]. This price aspect of value was already highlighted by famous thinkers such as Karl Marx (1865/2010)[11] and David Ricardo (1888).[12]

Additional definitions relate to value as "something (such as a principle or quality) intrinsically valuable or desirable"[13]. Kant (1785) already emphasized an intrinsic component of value related to human dignity and, therefore, not part of a market transaction, compared to price as an extrinsic characteristic for which a market exists.[14]

Nevertheless, how is it possible to narrow down the concept of value so that it becomes understandable and measurable within the context of organizational legitimacy? A structure for the logic of value, developed by Iwin (1975), brings more clarity to the term and its existing components. Relating to philosophical remarks and the term axiology[15], the author divided value into four components:

---

[7] see Mulgan et al. (2019); Talbot (2009).

[8] Bilolo (2018); Meynhardt et al. (2017); Meynhardt and Bäro (2019).

[9] Merriam-Webster (n.d.-b).

[10] Merriam-Webster (n.d.-b).

[11] Marx (1865).

[12] Ricardo and McCulloch (1888).

[13] Merriam-Webster (n.d.-b).

[14] see Kant (1785); as cited in Bilolo (2018), p. 82.

[15] The term was first mentioned by Paul Lapie (1902) in his publication called "Pour la Raison". Together with von Hartmann's (1908) "Grundriss der Axiologie oder Wertwägungslehre" both authors tried to define a theory of value.

the subject, an object, the basis, and the character of evaluation. The first, the subject, refers to any entity that is evaluating the object with the help of a value statement. Second, objects can be concrete or abstract forms, such as actions, measures, or policies. The third component, the basis, exemplifies the core nature of value since it represents the "forces" (e.g., needs, emotions, opinions, norms, ideals) initiating the evaluation"[16]. The character of evaluation as the fourth component deals with how values can be measured on a certain scale.[17]

The value components stress the differentiation between who values, what is valued, and how something is valued or evaluated. Discussions arose, and different approaches and schools of thought evolved around the core nature of value and whether it is more a matter of subjective evaluation and interpretation versus an objective value that independently exists for an object measured, communicated by its character of evaluation. Based on existing research, the objective view postulates that value is part of the object.[18] The subjective view focuses on the individual and bases their valuations on values.[19] The philosophic and contextual approach toward the definition of value merges these two views without judging what is wrong or right. This view states that besides these two opposing views, "value resides in the relation between the two, and does not exist apart from them"[20]. The interpretation of this view can be summarized in the sense that discussions about value would not exist if the subject would not see anything worth in the object and if a subject would not value the object.

Consequently, by stressing the relationship component of value, as public value theory does, value can be understood as the quality and "the result of a relationship"[21]. A relationship brings along two different stances: first, a subject that is responsible for the valuing of, second, a valued object. The transformation from a subjective evaluation to an objective value happens when a certain concordance exists regarding the evaluation.[22]

Nevertheless, what builds the reference point, the basis for this relationship? According to Menger (1871/2007), value is generated by "a judgement made by economizing individuals about the importance their command of the things has for

[16] Iwin (1975), pp. 49–55, as cited in Meynhardt (2009), p. 200.

[17] Meynhardt (2009).

[18] For further elaborations on value theory, its interpretations between objectivity and subjectivity, see Jalbert (1988); Rescher (1969).

[19] see Bilolo (2018); Guignon and Schnadelbach (1984).

[20] Alexander (1920), p. 302.

[21] Meynhardt (2009), p. 198; for the original source, please refer to Heyde (1926), pp. 46–47.

[22] see Meynhardt (2009), p. 199.

the maintenance of their lives and well-being"[23]. Value thereby focuses on basic human needs. The exploration of basic human needs as a crucial reference point for organizational legitimacy rests on public acceptance and public value.

## 5.2 Basic Needs as an Orientation and Reference Point for Public Value

Organizations such as enterprises can not neglect their social functions. Going further, Drucker (1954; 2001) postulated that organizations "are organs of society. They do not exist for their own sake, but to fulfill a specific social purpose and to satisfy a specific need of a society, a community, or individuals"[24].

From a corporate perspective, public value plays a crucial role when it comes to the evaluation of corporate actions that are related to the fulfillment of basic needs.[25] Understood as the "whats of the shoulds"[26], basic needs serve as measurement criteria and reference points to evaluate the difference between a perceived status quo and the desired, but mostly hypothetical, optimum.[27]

The origins of basic needs theory are rooted in the field of psychology. Epstein (1989) structured basic needs in his cognitive-experiential self-theory into four different needs that can be described "as a personal theory of reality"[28]. These four needs deal with psychological processes and human psychological and essential functions that include "assimilating of the data of reality within a relatively stable, coherent conceptual system; maintaining a favorable pleasure-pain balance over the anticipated future; maintaining relationships with significant others and maintaining a favorable self-esteem" [29].

As postulated by Epstein (1989)[30], these needs are a driving force, guiding individuals as their fulfillment is a key priority. The author called the needs universal, while the relationship between these needs might be dominated by a certain need

[23] Menger (1871), as cited in Bilolo (2018), p. 86.

[24] Drucker (2001), p. 963.

[25] see Meynhardt (2009); Meynhardt (2015).

[26] Meynhardt (2009), p. 201.

[27] Meynhardt (2009), p. 200.

[28] Epstein (1989), p. 8.; as cited in Meynhardt (2015), p. 152.

[29] Epstein1989, p. 8.; as cited in Meynhardt (2015), p. 152.

[30] Epstein's (1989) cognitive-experiential self-theory seems to reduce the nature of human beings to some limited functions. However, it is based on the valid assumption that the most complex, culturally mediated, and functionally autonomous values 'grow out' of basic dimensions (Hansch (1997), as cited in Meynhardt (2009), p. 202).

depending on the context. If needs are not satisfied, a negative impact on a person's conceptual system is likely.[31]

Meynhardt (2009), in a first step, took the self-theory by Epstein (1989) a step further by concentrating on four basic needs.[32] These four basic needs include the need for:

- positive self-evaluation
- maximizing pleasure and avoiding pain
- gaining control and coherence over one's conceptional system
- positive relationships

Transferring this concept to related actions, Meynhardt (2009) summarized that, first, motivations for a positive self-evaluation are, for example, building a relationship with the environment to have a specific reference for evaluating the self. Second, maximizing pleasure and avoiding pain involves getting to know positive and satisfactory experiences and positive feelings due to one's actions, while third, controlling the conceptual system means managing expectations of the environment and understanding its driving forces. The fourth basic need relates to positive relationships that include a balanced relatedness with the environment, such as being attached to a group and part of the public.

As a second step, the author acknowledged that basic needs are part of interrelated basic values. Following needs is a motivation oriented towards actions that can be categorized into four value dimensions. The connection between basic needs and the concept of value is rooted in the fact that individuals "'implicitly value' when fulfilling their needs"[33]. Meynhardt (2009) summarized the following four basic value dimensions:

- moral-ethical
- hedonistic-aesthetical
- utilitarian-instrumental
- political-social

Values and related actions are elaborated by Rokeach (1968), a famous psychologist, who highlighted that values function as criteria for evaluation and moral judgment

[31] see see Bilolo (2018); A. Epstein 1989; S. Epstein 2003.

[32] Meynhardt (2009), p. 203.

[33] Meynhardt (2009), p. 202.

by coining the term value as a "preferable mode of conduct or end-state of existence over other such alternatives"[34].

By considering this preferable mode of conduct that human beings strive for, the first dimension, the moral-ethical dimension, as one basic value, implies the perception and reflection of a compass for just and unjust actions, its embeddedness in a social context and the implications of decisions and actions on the matter of equality. Meynhardt (2015) spoke of a bar for an evaluation that represents "a perceived discrepancy between what he or she feels is appropriate or fair and the actual experience"[35]. In the context of organizations, the crucial question in that dimension is what kind of impact do organizations have on moral-ethical values, and how do they contribute to society?

The second dimension, the hedonistic-aesthetical one, relates to Epstein's (1989) pleasure and pain balance that highlights the need to make experiences with a positive connotation. It includes needs rooted in hedonism and certain preferences that are linked to the beauty or appearance of something. Here, the question in the organizational context is how individuals rate their experiences with organizations. Are they of a certain aesthetic quality, or does it make fun to get into an exchange with it, does it cause pain or create suffering?[36]

The utilitarian-instrumental aspect covers the third dimension, which triggers the efficiency and utility of an action or decision. Organizations offer products that customers and other stakeholders evaluate in terms of this dimension. People can decide what products they buy and how certain organizations help solve problems and conform to their world perception.[37] As Meynhardt (2015) acknowledged, money is part of this dimension. However, with a public value lens, money can only create public value if it is covered by a certain appreciation and an efficient benefit related to it.[38]

The fourth dimension, the political-social dimension as a power-related dimension, deals with the relationship aspect of public value. Relationships can be characterized by mutual support, loyalty, cooperation, inclusion, status, prejudice, and exclusion. This dimension is fueled by a certain belongingness and attachment to groups people strive for.[39]

---

[34] Rokeach (1968), p. 124, as cited in Bilolo (2018), p. 88.

[35] Meynhardt (2015), p. 153.

[36] see Meynhardt (2009), p. 209.; Meynhardt (2015), pp. 153–154.

[37] Meynhardt (2015).

[38] Meynhardt (2015).

[39] Meynhardt (2015), pp. 153–154.

The public value landscape by Meynhardt (2009), part of the following section, helps transfer the four basic value dimensions into a practical context.[40]

## 5.3 The Public Value Landscape and the Measurability of Public Value

How can individual basic needs turn into legitimacy criteria attributed to organizations? As Meynhardt (2015) described, "the public is inside"[41]. The crucial connection between individual basic needs and public value is rooted in individuals endowed with basic needs and their perception of the social environment and their experiences they share with the public. Public value, therefore, describes an abstracted reference point of a collective, representing the public, the society, and the community. Public value is not reduced to any individual characteristic of its members. Individuals and their needs are the source that turns public value into a value of the public, or as Meynhardt (2015) emphasized, "it always takes subjects to call a value into existence"[42].

These individual values can be a part of the public perspective by becoming what Dworkin (1978) called "impersonal preferences"[43]. In his statements, the author related, as an example, to environmental issues that societies and their citizens must face and counteract. Individual values become publicly accepted and a guiding principle for organizational legitimacy. They are demanded and become part of organizations' justifications, not a utilitarian benefit. Consequently, Sagoff (1982) stated that

> "many people believe that pollution should be strictly controlled, wilderness areas preserved, and the safety of workers assured not because these goals promise advantages to them but because they conform to their conscientious idea of what a good society stands for or obliges itself to do"[44].

The public value landscape by Meynhardt (2009), as shown in Fig. 5.1, helps to assess and measure issues and topics that relate to individual basic needs and their value dimensions.[45]

---

[40] see Meynhardt (2009), p. 208.

[41] Meynhardt (2015), p. 151.

[42] Meynhardt (2015), p. 151.

[43] Dworkin (1978), p. 127., as cited in Sagoff (1986), p. 310.

[44] Sagoff (1986), p. 311.

[45] see Meynhardt (2009), p. 208.

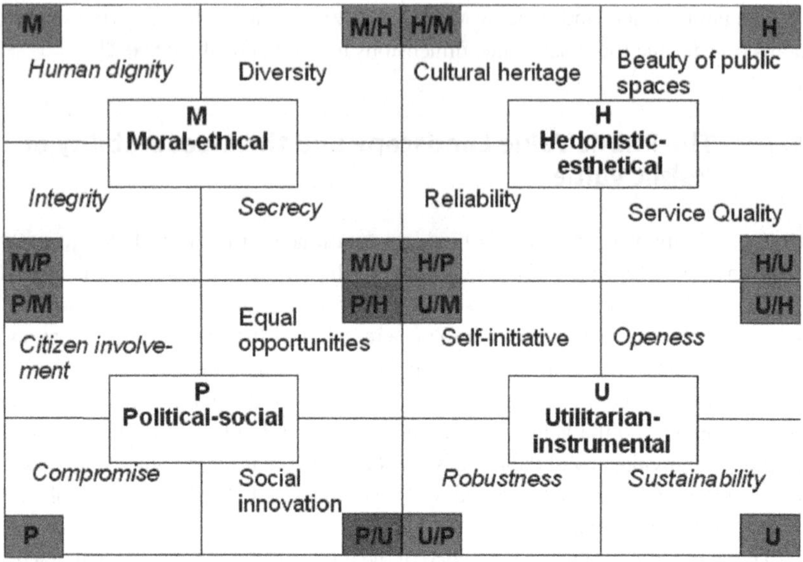


**Fig. 5.1** The Public Value Landscape (Meynhardt 2009)

The landscape supports empirical work such as the one in this dissertation in the classification of human basic needs that become part of the societal discourse and, thereby, a crucial aspect of organizational legitimacy. It must be acknowledged that Fig. 5.1 only shows a selection of possible topics that are assigned to the four basic value dimensions. The landscape supports the empirical investigation of this dissertation to allow the assessment and measurement of public value topic salience and the underlying issues.[46]

---

[46] Though representing a suitable classification framework in the context of basic needs and values, Meynhardt (2009) acknowledged the existing limitations of such a classification that range from a possible interpretation bias to a lack of more significant empirical evidence. However, at the exact moment, the author highlighted that for the dimensions and values of the landscape, "we would need to consider them to capture the 'full roundedness' of human nature"(Meynhardt (2009), p. 208).

## 5.4 Societal Discourse and Public Value

If values are not shared and corporate behavior does not meet societal expectations, organizational possibilities and growth are endangered to be limited due to a company's opposing views and actions being in confrontation with its stakeholders.[47] If and how companies interact with their stakeholders, the underlying issues of that relationship become a question of organizational legitimacy and the values that the relationship represents.[48] It emphasizes the importance of the maintenance of this relationship. Companies are thereby part of and nestled in the relationship with the individual and the collective and are asked to remain in good standing.[49]

Good for society thereby means being part of and positively impacting society and the social environment. Though knowing theoretically and with the help of the public value landscape how to measure public value and how to classify its underlying values that emerged out of individual basic needs, how do we know more about the substance of public value and what the public values?[50] What issues foster organizational legitimacy and what societal needs have to be on the radar of organizations?

Hirschman (1982) already acknowledged that societies fluctuate "between periods of intense preoccupation with public issues and of almost total concentration on individual improvement and private welfare goals"[51]. Thus, discourse plays a significant role in the societal detachment towards its issues since values are expected "to orient in two directions, that of discourse and that of overt action"[52]. Detecting societal discourse is a radar for public issues relevant to public value creation. It converts to the question of how organizations can stay tuned in societal discourse and how they can generate the knowledge to let actions follow.

In the interpretation of Drucker (2001), "managers must convert society's needs into opportunities for profitable business"[53]. Investments and revenue generation are necessary to grow and survive. However, at the same time, the public is a judge that decides if organizations survive based on their organizational legitimacy and contribution to public value.[54] Transferring a negative interpretation into a positive

---

[47] cf. Rüegg-Stürm and Grand (2019).

[48] see Bilolo (2018); She and Michelon (2019).

[49] cf. Meynhardt (2015), p. 148.

[50] Talbot (2011), as cited in Meynhardt (2015), p. 150.

[51] Hirschman (1982), p. 3., as cited in Meynhardt (2009), p. 200.

[52] Rescher (1982), p. 3., as cited in Meynhardt (2009), p. 199.

[53] Drucker (2001), p. 972.

[54] see Benington and Moore (2011); Meynhardt and Bäro (2019); M. H. Moore 2013.

one, the hypothetic consequence can result in a greater alignment of organizations with the public interest and stakeholders. Such an alignment increases the chance of generating public value. According to Drucker (2001), "business should hold a front seat in public discourse"[55] and become familiar with managing its public value.[56]

Focusing on the actions that create public value leads to an issue-related exploration of topics that dominate societal discourse. Stakeholder relationships exist "in order to find a common approach to an issue that affects them all, and that is too complex to be addressed effectively without collaboration"[57]. Saffer (2019) stated that "organizations and publics are not stakeholders to each other but stakeholders to the issues that brings them together"[58]. Issues and the mobilization of public interest shape stakeholder relationships of organizations,[59] whereas the values behind and attitudes towards these issues define the quality of relationships. Putting the issue, its actions, and the underlying interactions with stakeholders into focus enables the empirical analysis of public value and its value relevance.[60]

55 Bilolo (2018), p. 70.; see Drucker (2001).

56 see Meynhardt and Bäro (2019), p. 93.

57 Roloff (2008), p. 238.

58 Saffer (2019), p. 284.

59 Jeffery (2009), p. 17.

60 see Adler and Kwon (2002); Saffer (2019).

# Summary Theoretical Findings 6

Various publications and the work of dedicated researchers over different domains support this dissertation by providing a profound theoretical basis. The central theme of this dissertation dealing with the value relevance of stakeholder orientation and public value topic salience is rooted in the question of what drives and fuels organizational legitimacy. In a system-oriented view, organizations are even more complex when their environment is considered. Nevertheless, what reduces the complexity is knowledge generated by human beings and the use of communication for institutionalizing and legitimating specific patterns of conduct that exist and entail meaning for society.

The public and the social environment of organizations remains complex and characterized by a fluidity of discourse that shapes organizational legitimacy. Speaking about the outer context of organizations and recapturing the words of Hoffman and Ross (1999), who stated that "issues define what the field is"[1] puts the lens on the players that confront organizations with current issues and the substance of these issues.

Part of this environment are stakeholders, be it stakeholders from a social or financial background that exert expectations and, beyond that, are in a position to challenge organizational legitimacy that is crucial for organizational survival. Stakeholders are moving to the front seats when it comes to raising their voices to decide about organizational licenses to operate. Based on the theoretical framework by Mitchell et al. (1997), all three aspects are key to identifying relevant stakeholders: power, urgency, and legitimacy.[2] These aspects become crucial since the urgency is obvious in the context of the grand challenges that the world is facing. At the same time, the power of stakeholders to impact organizational legitimacy is

[1] Hoffman and Ross (1999), p. 352.

[2] Mitchell et al. (1997).



T. Heroth, *Corporate Sustainability Disclosure Matters*,
https://doi.org/10.1007/978-3-658-47653-3_6

increased, leading to a more and more unquestioned legitimacy of certain stakeholders. Stakeholder theory helps to broaden the scope and enrich the understanding of how companies are embedded and influenced by their social environment.[3] It further postulates that the better a company manages its relationships with various stakeholders, the better its performance will be, or as Drucker (2001) postulated, justification will only be possible if organizations are good for society.[4]

However, the orientation of organizations towards stakeholders, its prioritization, and the intensity of this connection is only one crucial aspect. Another question is what drives the contact, what issues dominate the exchange, and what challenges the organizational stakeholder relationship. To answer these questions with a focus on social issues that come into greater focus due to the developments towards sustainability that are presented in the following chapter, public value theory and the inclusion of basic needs provide a suitable framework. Individuals are endowed with basic needs. At the same time, basic needs become preferences and values that human beings expect to be impersonal and, thereby, part of the society that counts for all and not only for some. With the confrontation of issues and subjects, individuals "call a value into existence"[5]. The public value landscape enables a classification of basic needs and related values to allow their empirical investigation. Since the social environment is in flux, it is vital to put the societal discourse into focus to be able to draw further conclusions. We live in a time with a greater preoccupation with societal issues, not on a voluntary basis, but on an urgent and menacing one.

Therefore, detecting the importance of values in the organizational context by showing their financial and sustainability value relevance delivers essential aspects of public value creation and the management of organizations' outer context. Putting stakeholder theory and public value theory in the context of organizational legitimacy and into a relationship with the concept of sustainability that represents a priority on the corporate agenda is part of the following chapter.

[3] Hörisch et al. (2014).

[4] Drucker (2001), p. 965.

[5] Meynhardt (2015), p. 151.

# Sustainability and Materiality—Its Organizational Implications and Disclosure

# 7

Sustainability is a concept and a duty ranking high on the agenda of people and organizations in the 21st century. Going back to different domains in which a sustainable way of thinking and acting was proposed to preserve and protect nature, sustainability approaches evolved in areas such as hunting, fishery, and forestry. In the late 1990s, the first approaches to institutionalize sustainability developed. At that time, organizations were not yet confronted with more profound thoughts about reducing adverse effects on nature and the social environment. Companies did not have to react directly to the dynamics of stakeholder expectations and external pressure in a way that their organizational legitimacy was questioned. Due to its centrality in business, the development of the concept of sustainability, from its origins until today, is part of the following section to illustrate its meaning for organizations and corporate disclosure.

## 7.1 Sustainability—The Conceptual Historic Foundations

The foundations of the concept of sustainability go back to several thinkers and cultural traditions that evolved. The term sustainability with the German translation of *Nachhaltigkeit* goes back to the tradition of hunters who withhold some of the prey hunted to have something left for harder times, representing the so-called *Nachhalt* or what we are trying to sustain.[1] Banon Gomis et al. ( 2011) described that respect for nature was already a key part of attitudes towards life by early Confucian scholars and mentioned in Hebrew scriptures concerning a careful "treatment" of the earth.[2]

[1] see Reidel (2010), p. 102.

[2] see Bañon Gomis et al. (2011).



T. Heroth, *Corporate Sustainability Disclosure Matters*,
https://doi.org/10.1007/978-3-658-47653-3_7

With its link to the fishery domain, the term and concept of sustainability can also be explained by the name of a lake in South America called "Manchau gagog changau gagog chaugo gagog amaug", whose English translation is that we are fishing on our side, while you are fishing on your side. Nobody is fishing in the middle of the lake.[3] In the domain of forestry, in the early 18th century, Hans Carl von Carlowitz proclaimed a condition for the sustainable usage of wood and the treatment of forests by stating that "never harvesting more than what the forest yields in new growth"[4]. At that time, the focus was already on the regrowth and protection of certain species.[5]

In the 20th century, the first milestone towards sustainability with a scientific background was reached by a publication called *The Limits of Growth* published at the Massachusetts Institute of Technology (MIT) and supported by the Club of Rome. The document asked for a more systemic view of protecting the environment by considering the boundaries of the planet and the urgency of relevant actions.[6] Since then, various publications of researchers discussing the concept of sustainability received academic attention.[7] According to Marshall and Toffel (2005), already in the mid-1990s more than 100 definitions of sustainability existed.[8]

The first institutional approach to streamline and foster sustainability was part of a United Nations report from 1987 by the Brundlandt Commission called the Brundlandt Report that defined sustainability as "meeting present needs without compromising the chances of future generations to meet their needs"[9]. The quote tackles the central question of how can a better life be guaranteed without further pushing the degradation of nature. The report stresses the importance of the protection of the environment but also the so-called intra- and intergenerational equity or justice by emphasizing a fair distribution and equality among different generations and within one generation.

The World Summit of the UN that proclaimed the Rio Declaration in 1992 ignited the start of a frequent political format and platform to discuss sustainability-related topics. The declaration included sustainability as a principle for economic and political activities. The linkages between environment, economy, and social

---

[3] see Wördenweber (2017).

[4] Wiersum (1995), as cited in Kuhlman and Farrington (2010), p. 3437.

[5] von Carlowitz (1713), p. 105., as mentioned in Gabriel and Sailer (2023), p. 31.

[6] see Gabriel and Sailer (2023), pp. 31–32.

[7] see Banerjee (2003); Bañon Gomis et al. (2011); Gabriel and Sailer (2023); Kuhlman and Farrington (2010); Lankoski (2016); Leal Filho (2000); Marshall and Toffel (2005).

[8] Marshall and Toffel (2005), p. 673.

[9] World Commission and Development (1983), as cited Gabriel and Sailer (2023), p. 32.

issues highlighted in the declaration summarized that there is no economic progress without environmental protection while preserving the environment is only possible if social and economic matters are considered.[10]

The next groundbreaking summit that pushed the concept of sustainability was the Millennium Summit in New York in 2000, which focused on social and environmental issues. An agreement was published on goals to be achieved until 2015 that led to the Millennium Development Goals, including topics such as human rights, poverty reduction, peace, and the protection of the environment. Representing the succeeding goals that served as the first conclusion of the successes reached by the Millennium Development Goals and the gaps remaining to be closed, the SDGs were proclaimed by all UN Member States in 2015.[11] In another period of 15 years, the SDGs expect the commitment of the member countries, with a more substantial responsibility of developed countries and organizations, to follow the 17 SDGs and its subgoals.[12]

Under the term sustainability, the Brundlandt Report referred to economic, ecological, and social living conditions.[13] According to Gabriel (2023), economic sustainability takes the securing of income into account, while ecological sustainability focuses on the fact that human needs satisfaction is only possible in “a way that the capacity of the ecosystems surrounding us is not exceeded and biological diversity does not suffer”[14]. Social sustainability puts the lens on satisfying basic human needs. It further tackles social justice and security retention while aiming for equality.[15]

In line with that three-part approach, a model that gained attention focusing on the organizational context is the triple bottom line by Elkington (1997). Intending to come closer to certain concept metrics, especially in business, the author categorized a social, economic, and environmental bottom line to measure progress in the three categories.[16] Until today, this multi-dimensional view of sustainability is favored

---

[10] World Commission and Development (1983), as cited Gabriel and Sailer (2023).

[11] see World Commission and Development (1983), as cited Gabriel and Sailer (2023); Bebbington and Unerman (2018).

[12] A detailed view on the SDGs with its link to the public value landscape by Meynhardt (2009) is presented in section 13.6.4 in the empirical part of this dissertation.

[13] see Kuhlman and Farrington (2010), p. 3438; Gabriel and Sailer (2023), p. 32.

[14] Morelli (2011), p. 5.; Gabriel and Sailer (2023).

[15] Gabriel and Sailer (2023); Rasouli and Kumarasuriyar (2016).

[16] see Elkington (1997).

by many organizational researchers and practitioners who sometimes also refer to this view under the concept of people, planet, and profit.[17]

Regarding the triple bottom line, Marshall and Toffel (2005) stated that "organizations pursuing sustainability ought to make decisions based not only on economic returns but also on environmental protection and social justice"[18]. However, in the business context, the interpretation of sustainability and its definitions are still manifold. This is why, in the following sections, this dissertation concentrates on sustainability in business, focusing on its regulations and disclosure with a unique lens on its social dimensions.

## 7.2 Sustainability in the Organizational Business Context

Although the foundations of the concept of sustainability relate to countries and politics to channel measures towards a sustainable future and to counteract environmental and social issues, the contribution of business organizations in that context is apparent. It is business organizations that contribute most to carbon dioxide and other emissions while, at the same time, being a guarantor for jobs, social security, and technological innovations. In a corporate business context, sustainability can be driven by factors such as regulation, sustainable business models and strategies, financing sources and stakeholders, namely banks and investors, or pressures from customers, pressure groups, and society. Consequently, organizational legitimacy is at stake.

Apart from the triple bottom line by Elkington (1977), the term corporate social responsibility (CSR) found its way into business. According to Carroll (2008), the concept of CSR has its roots in the Industrial Revolution and found its way into the business context starting in 1930 with the beginning of the "corporate period"[19], when reflections about the social obligations of organizations began to evolve.[20] At that time, the term social responsibility (SR) was more often used than CSR, showing that it was more on the business people to be responsible, later replaced by companies dominating the business sector. In the 1950s, the first definition of social responsibility stated that social responsibility "refers to the obligations of

[17] see Palmer and Flanagan (2016).

[18] Marshall and Toffel (2005), p. 674.

[19] Eberstadt (1977).

[20] see Carroll (2008), p. 24.

businessmen to pursue those policies, to make those decisions, or to follow those lines of action which are desirable in terms of the objectives and values of our society"[21].

In the 1960s, SR definitions came close to what we today understand as sustainability with a lens on public values. As Frederick stated in 1960, "social responsibility in the analysis implies a public posture toward society's economic and human resources and a willingness to see that those resources are utilized for broad social ends and not simply for the narrowly circumscribed interests of private persons and firms"[22].

The integration of a stakeholder-oriented view was initiated by statements going back to the 1970s, when first publications and statements evolved about the multiplicity of interests that exist in business, not only accounting for the stockholders and its profits but also for a variety of groups with further specific interest.[23] At the beginning of the 21st century, CSR became a central focus of business globally. Pushed by more and more companies that reported voluntarily on CSR and by publications that proclaimed a new way of doing business that considers and creates value while proactively addressing stakeholders respectfully.[24]

Shortly before the start of a new century, in 1999, efforts to support sustainability in business received greater attention because the General Secretary of the UN at that time, Kofi Annan, demanded a new global economy by calling to the members of the World Economic Forum in Davos that "individually through your firms and collectively through your business associations—to embrace, support and enact a core set of values in the areas of human rights, labour standards, and environmental practices"[25].

One year later, the ambitions to better unite business with the goals of the UN were manifested by establishing the UN Global Compact.[26] Under the Global Compact and as an initiative between the UN and the Swiss Federal Department of Foreign Affairs of the Swiss Government, the report *Who Cares Wins* was published with the subtitle *Connecting Financial Markets to a Changing World.*[27] Although this report was, as the name says, addressed to financial markets, it related to research and investment procedures that shall consider environmental, social, and governance

---

[21] Bowen (n.d.), p. 6., as cited in Carroll (2008), p. 26.

[22] Frederick (1960), p. 60., as cited in Carroll (2008), p. 26.

[23] Johnson (1971), as cited in Carroll (2008), p. 26.

[24] see Carroll (2008); Kotler and Lee (2008); Perrini and Tencati (2006).

[25] United Nations (1999).

[26] see Nolan (2005), p. 446.

[27] United Nations Environment Programme Finance Initiative (2004).

concerns, which coined the acronym ESG that until today not only created new ratings and investment products with trillions US Dollars of asset value but also became a guiding principle in sustainability disclosure.[28]

As mentioned, definitions of sustainability in business or corporate sustainability vary according to its research and practical focus. Montiel and Delgado-Ceballos (2014) analyzed publication trends and the definitions evolving around the concept of corporate sustainability. Some authors limited the focus of corporate sustainability to a bidimensional or singular environmental management dimension.[29] Nevertheless, most publications referred to the three pillar of the triple bottom line and a clearer definition of its subtopics, such as the definition of Szekely and Knirsch (2005) who defined sustainability in business as

> "sustaining and expanding economic growth, shareholder value, prestige, corporate reputation, customer relationships, and the quality of products and services as well as adopting and pursuing ethical business practices, creating sustainable jobs, building value for all the stakeholders and attending the needs of the underserved"[30].

In this dissertation, the core theme lies in the social dimension, the values for stakeholders, and the existing basic needs. Coming from these needs that are linked with basic value dimensions, related stakeholders, and prominent topics and issues being part of societal discourse enables a focused empirical investigation. Therefore, sustainability can also be seen from a perspective focusing on its values. As Meynhardt clarified (2019), "the appeal of public value stems from its plurality of perspectives, since it also helps us address the widening gap between the 'facts of business life' and the public perception of what the economy is good for"[31], to avoid that the role of society and the public is not underestimated in the definitions of sustainability and to show its empirical evidence.

In a more practical context, recent regulatory efforts of governments, such as the Corporate Social Responsibility Directive (CSRD) of the EU, go beyond the abstract construct of society by acknowledging and naming social partners and communities as relevant and vital addressees and users of sustainability information

[28] Respecting the origins of the different concepts of the acronyms ESG and CSR and how they find its way into finance and business, since the focus of this dissertation lies on the disclosure of stakeholder orientation and social issues related to public value, the term sustainability disclosure or sustainability reporting are preferred and applied in the underlying context.

[29] cf. Montiel and Delgado-Ceballos (2014), p. 123.; Marshall and Toffel (2005).

[30] Székely and Knirsch (2005), p. 628, as cited in Montiel and Delgado-Ceballos (2014), p. 122.

[31] Meynhardt (2019), p. 8.

to be addressed by organizations.[32] Consequently, a factor that pushes a development towards a greater awareness of sustainability and the role of society and the public in the business context is regulation, with a focus on different geographical contexts and interpretations presented in the following section.

### 7.2.1 Current Developments in Sustainability Reporting Regulation and Related Standards

Besides its theoretical footings and existing definitions and approaches, sustainability relates to corporate actions and obligations that many organizations are globally confronted with. Regulatory standards and directives referring to the sustainability of organizations have increased severely in recent years. Organizations are challenged by how they do business and how their actions are externally perceived and reported. Governments of various countries and geographies take serious steps to counteract organizational actions that are not sustainable, passing more explicit and more binding directives and resolutions that lead to organizational obligations in the field of sustainability.

These regulations focus either on the investor or the corporate side.[33] The goal of such regulations is to find universal agreement, enabling companies with clear guidance on what to and how to report and investors to have a comparable basis for evaluating the sustainability performance of companies and portfolios.[34] Also, other stakeholders such as banks, civil organizations such as NGOs, and society are often digging in the dark to evaluate companies and potential sustainability concerns. However, there is a development toward more sustainability transparency pushed by sustainability regulations. To enhance the transparency of sustainability information, according to Cupertino et al. (2022), voluntary motivations to mandatory regulations and obligations for companies to disclose such information "could

[32] European Union (2023).

[33] Under the ESG pillars, the goals of regulations focusing on investors intend to guide investments in a sustainable direction. For regulations focusing on the investors' perspective, who stated that sustainable policy interventions are increasing significantly, emphasizing that 96% of all sustainable policies in the PRI database were passed since 2000 (United Nations Principles for Responsible Investment (2022)). In the EU, an important regulation with a clear focus on investors and financial services and their sustainability disclosure is the Sustainable Finance Disclosure Regulation (SFDR), whose main provisions under the regulation 2019/2088 have been applied since 2021 (European Commission (2023)).

[34] Lokuwaduge and De Silva (2022), p. 1121.

stimulate companies to become more socially responsible and pursue better ESG performance"[35].

The EU is a forerunner in the development and publication of sustainability regulations. With a focus on corporate obligations to disclose sustainability information driven by EU legislation, the following section details the recent regulatory sustainability regime.

#### 7.2.1.1 European Sustainability Reporting Legislation

The EU is trying to achieve the goal of no net greenhouse gas emissions until 2050, fostering economic growth independent from resource use while taking everyone on board and leaving no one behind.[36] These goals, being part of the European Green Deal, were presented in 2019. One of the focal points of these measures relates to the sustainability disclosure of companies. These transparency efforts aim to provide decision-makers with indications of how sustainable a company or its products are "behaving".

A significant impetus for EU regulations focusing on corporate sustainability was the Directive 2104/95 on non-financial disclosure and diversity (NFRD) published in 2014 that focused on corporate disclosure containing non-financial statements of "at least environmental matters, social and employee-related matters, respect for human rights, anti-corruption and bribery matters"[37]. The regulation relates to large public undertakings with an average number of over 500 employees. According to LaTorre et al. (2020), about 6,000 companies in the EU have been required to publish non-financial statements that contain the mentioned information.[38] According to KPMG ( 2023), the NFDR covered approximately 11,700 companies and groups across the EU.[39]

In 2019, the EU Commission aimed to review the NFDR, which led to the proposal of the Corporate Social Responsibility Directive (CSRD), Directive 2022/2464, whose aim is "to revise and strengthen the existing rules introduced by the Non-Financial Reporting Directive (NFRD), and to bring—over time—sustainability reporting on a par with financial reporting"[40]. In 2023, the CSRD went into force with the first obligations of companies to report on topics covered by the directive starting in the 2024 financial year. The CSRD classifies information to be disclosed

---

[35] Cupertino et al. (2022), as cited in Cicchiello et al. (2023), p. 1122.

[36] European Commission (2023); Koumbarakis et al. (2023).

[37] European Union (2014), p. 2.

[38] La Torre et al. (2020).

[39] KPMG (2023).

[40] European Parliament (2023).

on environmental, social, and governance factors. While the environmental factors focus on climate change mitigation, resource use, and pollution, the governance factors include risk management and business ethics. The social and human rights matters include factors such as equality and diversity, working conditions, and points related to the Bill of Human Rights and UN Human Rights Convention such as respect for freedom and human rights.[41] According to KPMG ( 2023), the number of companies affected has risen to approximately 49,000 companies EU-wide since the criteria for mandatory reporting were amended to all large companies[42] and all listed companies with 250 employees, EUR 40 million turnover, or EUR 20 million in total assets.[43]

Although a particular impetus for sustainability regulation can be attributed to the EU, other geographies and countries also follow a stricter sustainability reporting agenda that organizations must follow. Examples of the US, China, and Switzerland are presented in the following section.[44]

#### 7.2.1.2 Regulatory Developments of Sustainability Reporting in Other Geographies

In the U.S., sustainability regulation is very much driven by the environmental dimension with a significant focus on sustainable investments related to ESG concerns. For example, in 2022 and for the year 2023, the U.S. Securities and Exchange Commission (SEC) issued and released a climate disclosure proposal that shall consider the suggestions of the Task Force on Climate-Related Financial Disclosures (TCFD) of the Financial Stability Board (FSB) for companies, which are subject of frequent reporting requirements such as Form 10-K and Form 10-Q reports for public companies.[45] In addition, the United States Department of Labor, as the regulator of employee benefit plans in the private sector, issued rules in 2022 to include ESG factors in the investment decisions of fiduciaries.[46]

---

[41] European Union (2023).

[42] "Large companies refers to companies not established in the EU that are listed on EU regulated markets and EU subsidiaries of non-EU companies"(Deloitte (2022)).

[43] Deloitte (2022); KPMG (2023).

[44] It is obvious that a selection of regulatory developments in the mentioned countries can only partially represent a complete picture of sustainability regulations on a global scale. However, the selection of the four different jurisdictions does not only represent most companies of the underlying data sample but also a representative picture of the global landscape of sustainability reporting.

[45] see Linklaters (2022).

[46] ICLG (n.d.).

Though it is a dominant player in sustainable investing, there is an apparent lack of more detailed ESG reporting and disclosure regulations in the United States.[47] Cicchiello et al. (2023) quoted the statement by the acting chair of the SEC, Allison Lee, in 2021, stating that there is an "urgent need to increase and improve the quantity and quality of non-financial disclosure and its intention to develop a comprehensive framework that produces consistent, comparable and reliable climate-related and ESG disclosures"[48].

The World Business Council for Sustainable Development (WBCSD) identified 211 reporting provisions for external reporting, whereas only 27% were mandatory. The U.S. Environmental Protection Agency pushes with a strong focus on the environmental pillar while social issues are developing but still lagging.[49]

US guidances that exist for companies include broader ESG risks to be disclosed by companies under SEC supervision in the Management's Discussion and Analysis (MD&A) or Item 303 of the required 10-K annual report calling for the disclosure of "those key variables and other qualitative and quantitative factors that are peculiar to and necessary for an understanding and evaluation"[50] of a company. Apart from these broader regulations mainly affecting all at least public companies, institutions such as the exchange NASDAQ have introduced regulations for companies listed on the exchange that are required to disclose information on board diversity and the gender-related LGBTQ+ status of a company's board.[51]

In China, the Shenzhen Stock Exchange 2006 spoke in favor of a voluntary sustainability disclosure of the companies on the exchange; the Shanghai Stock Exchange implemented mandatory disclosure of sustainability reports of their respective companies.[52] A push for ESG regulation in the country was due to the revision of the Code of Corporate Governance by the China Securities Regulatory Commission (CSRC) in 2018 and new disclosure rules for listed companies concerning their annual and semi-annual reports. Listed companies must publish information on their stakeholders and social and environmental issues.[53] Disclosure

[47] It has to be mentioned that laws related to sustainability and ESG in the United States are under scrutiny, especially on the state level. State lawmakers introduced approximately 165 "anti-ESG" proposals in 37 states, while 28 of these provisions were approved. These bills shall restrict the usage of ESG data or set penalties if specific industries are excluded due to ESG investment criteria (United Nations Principles for Responsible Investment (2023)).

[48] Cicchiello et al. (2023), p. 1122.

[49] World Business Council for Sustainable Development (2018), p. 4.

[50] ICLG (n.d.).

[51] ICLG (n.d.).

[52] see Yuan et al. (2022).

[53] Lee (2021).

regulations in China are pushed by the Chinese Environmental Protection Law that was revised in 2022 and to laws related to the environment dimension.[54] Furthermore, the China Enterprise Reform and Development Society (CERDS) published detailed ESG criteria that companies shall disclose, though still voluntarily.[55]

In Switzerland, in 2022, the parliament implemented new provisions for the year 2023 on non-financial reporting as an indirect counterproposal after an initiative of Swiss voters under the name responsible business initiative (RBI) was rejected by Swiss polls. Being now part of the Swiss Code of Obligations (CO), the obligations for companies include non-financial reporting obligations under article 964a-964c CO that are mandatory for companies of public interests above 500 employees when exceeding either a threshold of CHF 20 million CHF as a balance sheets total or revenues out of sales of CHF 40 million.[56] Information to be disclosed by the mentioned companies shall "cover environmental matters, in particular, CO2 goals, social issues, employee-related issues, respect for human rights and combating corruption"[57].

Linked to this obligation is the article Art. 964j-964l CO covering special due diligence requirements and the obligation on a transparent disclosure of information related to conflict minerals and child labor. This article counts for all undertakings in Switzerland related to minerals or where products or services are offered "to which there is a reasonable suspicion that they have been manufactured or provided using child labour"[58].

What all sustainable regulations, standards, and guidelines have in common is the necessity to position towards a concept of materiality that is followed. However, differences exist globally in the understanding of whom companies are accountable to in their sustainability disclosure and, with this, what kind of non-financial information is material to report. Considerations of different standard setters and institutions regarding the concept of materiality are part of the following section.

[54] Gao (2023).

[55] see Chan (2022).

[56] Swiss Sustainable Finance (2023).

[57] Swiss Sustainable Finance (2023).

[58] Swiss Sustainable Finance (2023).

## 7.3 Considerations of Materiality with Reference to Existing Reporting Standards

Mainly, all regulations in the field of sustainability related to organizations do, to different extents, refer to the concept of materiality. However, the concept is more complex than it might be perceived at first glance. Since almost all bigger companies are obliged to report their financials with the help of at least a yearly report and accounting standards exist to structure and guide financial reporting, the mentioned increasing reporting regimes for non-financial information question the existing concept of materiality. The importance of this subject is described by the fact that an increasing number of companies, be it due to obligations or voluntarily, disclose sustainability reports and non-financial information. In 2019, over 90% of S&P 500 firms published information on sustainability. This is a sharp increase compared to 2011, when the number of companies providing insights on ESG was less than 20%.[59]

Financial and non-financial reporting and its related reporting obligations both draw on the concept of materiality. However, two different principles exist that influence the understanding of what is material concerning sustainability information and what, therefore, should be reported. While referring to the origins of the concept and its definitions, both principles are presented in the next section, after a short introduction into the concept of materiality and its origins.

### 7.3.1 The Concept of Materiality and its Origins

The concept of materiality is familiar to companies and academic researchers in business and accounting. Bernstein, back in the days in 1967, presented a definition of materiality by referring to the SEC's Regulation S-X:

> "The term 'material' when used to qualify a requirement for the furnishing of information as to any subject, limits the information required to those matters as to which an average prudent investor ought reasonably to be informed before purchasing the security registered"[60].

In the United States, besides the SEC, different institutions such as the U.S. Supreme Court and the Financial Accounting Standards Board were included in the discussion

[59] Governance and Accountability Institute (2020).

[60] Bernstein (1967), p. 88.

about the term and its application.[61] In 1976, the U.S. Supreme Court decided that materiality is present if "a 'substantial likelihood' [exists] that disclosure of the fact would be viewed by the 'reasonable investor' as having altered the 'total mix' of information made available"[62]. Not without criticisms of several authors characterizing the definition as being murky while causing difficulties for companies to decide what to disclose and what not, the definition sustained.[63]

After several years, concerns about materiality evolved again, asking for additional research in that field and attracting practitioners, standard setters, and academia.[64] Messier et al. (2005) spoke of renewed concerns of the SEC and the Auditing Standards Board (ASB).[65] In a time where concerns are discussed with a focus on financial information and in the context of sustainable reporting, non-financial information found its way into an increasing amount of governmental regulations and standards, thereby bringing in a new classifier and new institutions with definitory power over the concept of materiality. The different principles of materiality are illustrated in the next sections.

### 7.3.2 The Principle of Financial Materiality in Corporate Disclosure

It remains the question of what attributes of corporate disclosure influence corporate value and its performance metrics and how materiality is perceived in this context. Therefore, it is crucial to understand better the concept of materiality with its different interpretations, its existing research, and its relationship with the concept of value and to enrich it with further empirical explorations. As the EU in paragraph 8 of the CSRD emphasized, "many stakeholders consider the term 'non-financial' to be inaccurate, in particular because it implies that the information in question has no financial relevance. Increasingly, however, such information does have financial relevance"[66]. This leads to three key questions. First, is sustainability, its performance, and information disclosed in sustainability reports financially relevant? Second, what kind of information is relevant for financial materiality considerations? Third, as shown in the subsequent section, can materiality only be interpreted in a financial sense?

---

[61] see Jebe (2019).

[62] Court (1976), as cited in Jebe (2019), p. 6–7.

[63] Court (1976), as cited in Jebe (2019); Schulzke and Berger-Walliser (2017), p. 8.

[64] see Edgley (2014).

[65] Messier et al. (2005).

[66] European Union (2023).

The SEC interpretates materiality from a financial perspective by limiting the information audience to the "average prudent investor"[67] that, as the definition of financial materiality by the European Union as part of the NFDR contributed, "is necessary for an understanding of the company's development, performance and position"[68].

Recent studies that explore information being part of corporate disclosure and its financial materiality are manifold.[69] These papers have a heterogeneous view on how to measure and define sustainability and its related performance indicators on the one side[70] and which financial performance metrics to be considered on the other side. In many studies, sustainability ratings[71] and sustainability and integrated disclosure quality scores[72] act as a proxy for sustainability performance, whereas accounting-based metrics such as Return on Assets (ROA) and Equity or market-related metrics such as Tobin's Q represent financial performance metrics.[73] In addition, several studies showed an effect of the quality of corporate reporting and sustainability disclosure on financial materiality by emphasizing a reduction of a company's cost of equity[74] and the cost of debt.[75]

However, a high-level view on sustainability and its link with financial performance is problematic since "relationships between ESG and financial performance are difficult to generalize without a precise understanding of the underlying ESG components and concepts and how these components are combined"[76]. Besides, sustainability ratings are contested in research due to their lack of comparability concerning the weightings and scopes of existing providers.[77]

Going from a high-level view on sustainability into the analysis of concrete sustainability constructs and ESG components is key for answering the second question. A holistic review of sustainability information that is financially material is hard to imagine within the bounds of possibility. It is also not in scope for this dissertation

[67] Bernstein (1967), p. 88.

[68] Baumueller and Sopp (2022), p. 9.

[69] cf. Fatemi et al. (2018), Friede et al. (2015), Giese et al. (2019), Landi and Sciarelli (2019), Ng and Rezaee (2015), Velte (2017).

[70] some papers refer to the acronym ESG or CSR.

[71] cf. Giese et al. (2019), Landi and Sciarelli (2019), Velte (2017).

[72] cf. Churet and Eccles (2014), Fatemi et al. (2018).

[73] cf. Velte (2017).

[74] cf. Dhaliwal et al. (2011); El Ghoul et al. (2011).

[75] cf. Eliwa et al. (2021); Raimo et al. (2021).

[76] Giese et al. (2019), p. 3.

[77] Berg et al. (2019), p. 5, Berg et al. (2021).

to provide a holistic review of sustainability information that is financially material. However, it is of a clear focus to learn more about the value relevance of the constructs of stakeholder orientation and public value topic salience and thereby contribute to the understanding of the concept of materiality.

From a stakeholder theory perspective, Freudenreich et al. (2020) emphasized that it is the relationships that create a value creation network between companies and stakeholders that guarantee an operative viability of a company.[78] Velte (2017) supported this viewpoint by stating that the origin of a good sustainability and financial performance should be a successful stakeholder management.[79] Stakeholder theory further suggests that the better a firm manages its relationship with various stakeholders, the better will be its performance since business itself is "a set of relationships among groups that have a stake in the activities that make up the business"[80]. Therefore, by moving from high-level analysis to a more granular empirical investigation of sustainability components by linking stakeholders being addressed in corporate reporting with an analysis of its financial value relevance supports the detection of a possible material factor, which is stakeholder orientation.

The analysis of public value topic salience has the same focus. With the help of the theoretical contributions, the construct can be identified and isolated. As shown in the second part of the empirical analysis, topics that evolve in societal discourse are analyzed with the help of frameworks such as the SDGs and the public value landscape. As in the analysis of stakeholder orientation, the cost of capital, including additional control variables, serves as a measure to identify the financial materiality of this concrete construct in focus to find the answer to the second question mentioned above. However, the third question opens the discussion about a non-financial interpretation of materiality that leads to the principle of double materiality that is presented in the following subsection.

### 7.3.3 The Principle of Double Materiality in Corporate Disclosure

Apart from the SEC's definition of materiality, which refers to financial materiality, the principle of double materiality aroused the attention of standard setters, governmental bodies, practitioners, and researchers. With its *Fitness Check on the EU Framework for Public Reporting by Companies* in 2019, the EU Commission replied to concerns that characterized the applied materiality definition as "vague and

78 Freudenreich et al. (2020), p. 5.; cf. Freeman (2010).

79 Velte (2017), p. 169.

80 Parmar et al. (2010), p. 5.; Freeman (1984); Walsh (2005).

unsatisfactory"[81], questioning the distinction between financial and non-financial performance. These concerns led to the publication of new guidelines for non-financial reporting (2019/C 209/01) for climate-related information that referred to the principle of double materiality.[82] For the EU, double materiality includes the definition mentioned above of financial materiality or single materiality, while it additionally refers to information "necessary for an understanding of the external impacts of the company"[83]. The definition thereby relates to information that is "of most interest to citizens, consumers, employees, business partners, communities and civil society organizations"[84]. According to Baumueller and Sopp (2022) after analyzing the respective guideline and the definition of double materiality, it was still subject to criticism due to its nonbinding character and its lack of clear and precise differentiation between non-financial and financial information.[85]

The goal of mandating the European Lab of the European Financial Reporting Advisory Group (EFRAG) in 2020 was to ensure harmonization and consistency of corporate reporting. Double materiality was further in focus to be implemented into the reporting process. The new definition proposed by EFRAG, that was accepted by the EU Commission and applied in the issuance of the CSRD,

> "require reporting not only on information to the extent necessary for an understanding of the undertaking's development, performance and position, but also on information necessary for an understanding of the impact of the undertaking's activities on environmental, social and employee matters, respect for human rights, anti-corruption and bribery matters"[86].

The focused goals of the principle of double materiality are, first, to improve and foster stakeholder engagement.[87] According to several authors, stakeholder engagement under the concept of double-materiality "contributes to diverse and reciprocal accountability relationships between the organisations, their stakeholders, and the wider society and enables discussions and evaluations on sustainable development"[88].

[81] Baumueller and Sopp (2022), p. 18.

[82] Baumueller and Sopp (2022).

[83] EURLex (2019), p. 4.

[84] EURLex (2019), p. 4.

[85] see Baumueller and Sopp (2022); Sopp and Baumueller (2019); Korca and Costa (2020).

[86] European Union (2023).

[87] Puroila and Mäkelä (2019).

[88] Adams et al. (2021), p. 6; see Brown and Jesse (2015); Cooper and Morgan (2013); Puroila and Mäkelä (2019).

Baumueller and Sopp (2022) emphasized that larger volumes of accounting information can be expected, resulting in an information overload or the necessity to structure information more closely related to the stakeholders in focus. This increasing information requirements, especially such information that is not related to financial materiality, does, according to the authors, also incur higher costs for organizations and thereby question the related benefits.[89] Researchers with a focus on the outcomes of sustainability reporting emphasized that the current process for the determination of sustainable issues that are material needs to be more structured and needs an objective view. Speaking of a "credibility gap"[90] of sustainability information that is published with the help of sustainability disclosure, the authors highlighted that often, the communication of positive aspects covers negative ones, providing false and misleading information to stakeholders.[91]

Criticisms and different viewpoints regarding the understanding and application of the concept of materiality also affect sustainability standard setting and its related institutions, as illustrated in the following subsection.

### 7.3.4 The Landscape of Sustainability Reporting Standards—Its Standard Setters and Materiality Approaches

Facing difficulties on the way to finding reasonable approaches towards materiality, the leading global standard setters set different priorities concerning the definition of material information and the relevant audience of sustainability reporting.

In the United States, in 2022, a consolidation process of institutions resulted in the integration of the Value Reporting Foundation into the International Financial Reporting Standards (IFRS). The IFRS, a widely accepted international reporting standard, is issued by the IFRS Foundation and the International Accounting Standards Board (IASB). The Value Reporting Foundation itself was established due to a consolidation of the Sustainability Accounting Standards Board (SASB) and the International Integrated Reporting Council (IIRC) in 2021. While now part of the IFRS Foundation, the Value Reporting Foundation with the SASB Standards "serve as a key starting point for the development of the IFRS Sustainability Disclosure Standards, while the Integrated Reporting Framework provides connectivity

[89] Baumueller and Sopp (2022), pp. 22–23.

[90] Dando and Swift (2003).

[91] see Adams et al. (2021), p. 7; Adams (2004); Beske et al. (2020); Dando and Swift (2003); Knebel and Seele (2015).

between financial statements and sustainability-related financial disclosures”[92]. As a new vehicle of the IFRS Foundation, the International Sustainability Standards Board (ISSB) will represent the basic guidelines for sustainability disclosures for capital markets.[93]

According to Chiu (2022), with its main focus on investors and financial materiality, SASB standards are an appealing point of reference for investors and companies, while the mentioned consolidation further pushed the focus on single materiality. SASB standards are an industry-specific approach that refers to 11 sectors subdivided into 77 sectors with a range of two and a maximum of 11 material topics per sector.[94] Subjects are defined out of 26 generic sustainability topics that relate to the environmental and social dimension and topics related to social and human capital, leadership and governance, and business models and innovation. An industry classification is conducted due to the conviction that financial materiality is industry-specific.[95]

Several authors spoke of a clear inclination in the United States towards including single or financial materiality in SEC disclosure regulations.[96] Alexander (2022) emphasized that the goal of the ISSB is to have a certain uniformity to sustainability reporting compared with the characteristics of financial reporting that market participants concern and support. The author further acknowledged a clear rejection of the ISSB of the principle of double materiality. However, information that goes beyond single materiality is increasingly expected by investors for decision-making.[97] In addition, to come closer to a uniform approach towards materiality on a global scale, discussions started about requests that the ISSB should, as the European Banking Federation acknowledged, “consider criteria to target a wider set of stakeholders and move into the direction of double materiality”[98]. At the beginning of 2023, the ISSB voted to relate to standards of the Global Reporting Initiative and ESRS in its general sustainability-related disclosures (IFRS S1).[99]

In the EU, as the previous section and statement show, there is a great affiliation and motivation of EU bodies to foster the principle of double materiality in sustainability reporting. However, this focused approach towards double materiality does

---

[92] International Financial Regulation Standards (2022).

[93] International Financial Regulation Standards (2022).

[94] Consolandi et al. (2020), p. 3.

[95] see Consolandi et al. (2020); Eccles et al. (2012); Eccles and Serafeim (2013).

[96] see Chiu (2022) Harper Ho (2020); Saad and Strauss (2020).

[97] Alexander (2022).

[98] Verney (2023).

[99] Leng et al. (2023).

not mean that there is undisputed consensus among the leading institutions in the EU with a specific definitory power of what the principle means and how it should be assessed and disclosed by organizations with the help of sustainability reporting.

EFRAG, as the leading body responsible for the establishment and issuance of the ESRS, drafted its double materiality guidelines for standard-setting as a working paper in 2022 and an additional implementation guidance for the materiality assessment in 2023. The organization defined materiality as "the criterion for inclusion of specific information in corporate reports"[100] by further stating that it should reflect the significance of the information with its context and "its capacity to meet the needs and expectations of the stakeholders of an undertaking and of the undertaking itself, allowing for proper decision-making, and more generally [...] the needs for transparency corresponding to the public interest"[101].

Thereby, EFRAG (2023)[102] focuses on an approach that includes financial materiality and impact materiality of sustainable matters.[103]

According to paragraph 49 of the ESRS, financial materiality is defined with a direct reference to relevant financial metrics with the quote that includes

> "a financial perspective if it triggers or could reasonably be expected to trigger material financial effects on the undertaking. This is the case when a sustainability matter generates or may generate risks or opportunities that have a material influence, or could reasonably be expected to have a material influence, on the undertaking's development, financial position, financial performance, cash flows, access to finance or cost of capital over the short-, medium- or long-term"[104].

While the approach towards impact materiality that is stated in paragraph 43 of the ESRS follows that sustainability disclosure requires "an impact perspective when it pertains to the undertaking's material actual or potential, positive or negative impacts on people or the environment over the short-, medium- and long-term"[105]. Prior to being able to report on material topics, material assessments have to be conducted by companies being required by EFRAG. Material assessments shall

---

[100] EFRAG (2022a), p. 4.

[101] EFRAG (2022a), p. 4.

[102] EFRAG (2023) stated that the underlying publication is subject to change and that it "forms part of an early stage of the development of a potential EFRAG position. Consequently, the paper does not represent the official views of EFRAG"(EFRAG (2023)). However, it reveals crucial information about the positioning of EFRAG towards the concept of materiality.

[103] EFRAG (2023).

[104] EFRAG (2023), p. 8.

[105] EFRAG (2023), p. 8.

include "all material impacts, risks and opportunities of the undertaking related to a topic, sub-topic or sub-sub-topic […], and subsequently, the material information [… ] to be disclosed"[106]. EFRAG ( 2023) emphasized that materiality assessments shall be conducted based on the severity of an issue, including the scale, scope, and remediability as well as its likelihood.[107]

Besides EFRAG and its standards, GRI shows its stance in defining global sustainability reporting standards. Though GRI, with its first GRI guidelines published in 2000, was founded in Boston in 1997, it is closely affiliated with the approach of the EU towards sustainability reporting. Relocated to Amsterdam in 2022, GRI is a non-profit organization that, according to a study by KPMG in 2022, publishes sustainability reporting standards that are most applied globally.[108] According to Knebel and Seele (2015), "GRI as an NGO is perceived as a professional answer to provide guidance on economic, social, and environmental issues, as it provides guidance and performance indicators for companies on what to report"[109]. After the issuance of four general guidelines in the years 2000–2013 (G1–G4), GRI launched its sector program with a focus on 44 different industries in 2019.[110] In addition, in 2015, it decided to transmit the G4 Guidelines into the sustainability reporting standards, the GRI Standards. According to GRI (2015), this decision is part of its "move to become a standard setter"[111]. Part of this decision was the establishment of the Global Sustainability Standards Board (GSSB) to implement an additional and separate governance structure.[112]

When it comes to the concept of materiality, GRI follows a clear path of a multi-stakeholder approach that follows the principle of double materiality with its definition that was revised in 2020, stating that "the organization prioritizes reporting on those topics that reflect its most significant impacts on the economy, environment, and people, including impacts on human rights"[113].

In a GRI whitepaper that was composed together with academic researchers, the organization clearly emphasized its focus on double materiality due to the reason that a broader stakeholder perspective is vital to include since "different stakeholders have various, sometimes conflicting, views on material sustainable topics"[114].

---

[106] EFRAG (2023), p. 12.

[107] EFRAG (2023).

[108] KPMG (2022).

[109] Knebel and Seele (2015), p. 197.

[110] GRI (2023).

[111] GRI (2015).

[112] GRI (n.d.).

[113] GRI (2020a), p. 8.

[114] Adams et al. (2021), p. 6.; see Brown and Jesse (2015); Puroila and Mäkelä (2019).

Consequently, since different viewpoints and approaches towards double materiality exist, questions of convergence and applicability for organizations arose. To overcome these differences, EFRAG and GRI published a joint interoperability statement in 2023. The statements include that "existing GRI reporters will be well prepared to report under the ESRS. Entities reporting under ESRS are considered as reporting with reference to the GRI Standards and will therefore avoid the burden of multiple reporting"[115].

### 7.3.5 The Sustainable Development Goals as an Approach towards Double Materiality

A third approach towards sustainability reporting that combines the view of the presented standard setters in the EU that favor the principle of double over single materiality are the SDGs. According to the SDG Compass (2015) that acts as a guide for business concerning the SDGs and that resulted from a collaboration between the UN, GRI, and the WBCSD, "the SDGs are designed to rally a wide range of organizations, and shape priorities and aspirations for sustainable development efforts around a common framework"[116].

The following Fig. 7.1 shows the 17 SDGs that include environmental and social topics, and subjects related to economic health and cooperation.

Part of the 17 SDGs are 169 associated targets and 230 indicators that define tasks for their implementation. The SDGs are "ultimately aimed at achieving an 'ultimate end', a fair and prosperous, high quality of life that is equitably shared and sustainable"[117]. To aim for a realization of the SDGs in the organizational context, Szennay (2019) emphasized the role of businesses that with a Gross Domestic Product (GDP) comparable to states, companies "not only affect the societies of states belonging to their area of operation but they have a significant impact on the whole ecosystem of the planet"[118]. Furthermore, several authors highlight that the influences of the grand challenges that exist in the world transfer the understanding of business as a domain that does not accept a silo thinking of political and social versus economic, consequentially leading to a politicization of business.[119]

---

[115] GRI and EFRAG (2023).

[116] GRI (2015), p. 7.

[117] Costanza et al. (2016), as cited in Szennay et al. (2019), p. 3.

[118] Szennay et al. (2019), p. 5.

[119] Szennay et al. (2019); Wood et al. (2018).

**Fig. 7.1** The 17 United Nations Sustainable Development Goals (United Nations 2020)

The SDG Disclosure Recommendations defined materiality as "any information that is reasonably capable of making a difference to the conclusions drawn by stakeholders concerning the positive and negative impacts of the organisation on global achievement of the SDGs, and providers of finance concerning the ability of the organisation to create long term value for the organisation and society"[120].

Corporate reporting of SDGs increased vastly in the years 2017–2020.[121] The SDG Disclosure Recommendations emphasize its goal to affect the what and the how of business to be included in corporate disclosure. They include recommendations for disclosure that relate to the following central goals for disclosure: strategy, performance and targets, and governance and management. Considering materiality in all these aspects, strategy relates to the consideration of SDGs for strategic assessments of sustainable risks and opportunities of the business model and the value that is created for the organization itself and its stakeholders. Performance and targets highlight the disclosure of material financial and non-financial impacts. At the same time, governance includes the board's responsibilities and competencies for oversight of the defined materiality aspects and sustainable development issues

[120] Adams et al. (2020), p. 8.

[121] KPMG (2022).

that are defined by the management and influenced by managers, their personalities, attitudes[122] and openness for sustainability and its related topics.[123]

SDG disclosure represents an approach that is referenced by leading standards setters aiming to follow the principle of double materiality, such as EFRAG and GRI. By referring to the UN's 2030 Agenda for Sustainable Development in the published CSRD, the EU Commission set the goal that "European action for sustainability' links the SDGs to the Union policy framework to ensure that all Union actions and policy initiatives, both within and outside the Union, take those goals on board at the outset"[124]. In an explanatory note published by EFRAG (2022b), it is said that there is a clear reference in the first draft of the ESRS to the SDGs. The document structures the alignment of the ESRS towards ESG issues.[125]

GRI published its guidance and action platform for corporate reporting on SDGs in 2017, an initiative that was launched together with the UN Global Compact with a focus on supporting companies with SDG priority and target setting.[126] In the meantime, several publications and initiatives were published by GRI with the goal of linking the SDGs with GRI Standards and with the SDG Compass to enable the mapping of GRI standards with the single SDGs and its subgoals.[127]

Organizations face challenges implementing a double materiality approach and the SDGs into corporate reporting and strategies. Difficulties related to complexity issues, the number of goals and target indicators, and the potential or lack of interconnection with other reporting frameworks.[128]

Targeting SDGs in sustainability reporting is essential to incorporate the broader environment of the organization with its relevance and ability to affect its value.[129] According to Adams (2021), " 'value creation for organisations and society' in the SDGD Recommendations as one side of double-materiality has greater transformational potential than the EFRAG reference to 'financial' materiality"[130]. Chiu (2022) complemented, "embracing authentic double materiality requires new mindsets, resources and processes for measurement and evaluation, and would generate conflict

[122] see Carroll (1991).

[123] Adams et al. (2020), p. 12.

[124] European Union (2023).

[125] EFRAG (2022b), p. 12.

[126] GRI (2020a).

[127] see GRI (2020b).; For more information and research on the link between GRI Standards and the SDGs, see Szennay et al. (2019).

[128] see Ike et al. (2019); Szennay et al. (2019).

[129] Adams (2017).

[130] Adams et al. (2021), p. 5.

situations with single materiality that may not be easy to reconcile"[131]. Some authors spoke of a transformation necessary to implement the concept of sustainability holistically in organizations.[132] Hammond (2021) concluded that

> "rather than constituting a mere temporary management challenge, sustainability denotes a society's long-term response to socio-ecological phenomena that are complex, unpredictable, and constantly changing. Inasmuch as the causes of unsustainability are dynamic in nature, sustainability too must be understood as an open-ended, reflexive process, and transformability thus as a key component"[133].

Therefore, it is the question of empirical analysis to show potential evidence for the value relevance of stakeholder orientation and public value topic salience. Before illustrating the existing research gap, a summary summarizes the essential points of the underlying chapter.

## 7.4 Summary on Sustainability and Materiality

The previous chapter, with its focus on the concepts of sustainability and materiality, showed not only its close connection to the theoretical background of this dissertation but also the current developments, challenges, and different perceptions that exist for standard setters and organizations in shaping the process of sustainability reporting. Stricter sustainability regulations are developing in various countries and geographies around the globe, following the task to unpuzzle the complex concepts of sustainability and materiality for a clear definition of sustainability reporting requirements.

An inclination to traditions and the principle of single and financial materiality in countries such as the United States confronts with a push in the EU to support double materiality and the aspects of a multi-stakeholder approach that goes beyond the influence of sustainable aspects and issues of organizations on its profits but includes the impact of the business model on the planet, and the people.

Nevertheless, in the EU context, definitions of the principle of double materiality and its definitory power remain contested, asking for new mindsets necessary to realize and assess double materiality in business, as Chiu (2022) commented.[134] Not only for standard setters but also for organizations, it is helpful that further

[131] Chiu (2022), p. 102.

[132] see Abson et al. (2017); Chapin et al. (2010); Herrfahrdt-Pähle et al. (2020); Rees (2021).

[133] Hammond (2021), p. 174.

[134] see Chiu (2022).

research tries to verify and show the significance and usefulness for institutions and the people involved to be concerned with a more straightforward path toward sustainability reporting. This leads to the existing research gaps that shall contribute to bringing more light into the potential value relevance and materiality of stakeholder orientation and public value topic salience.

# Part II
# Empirical Investigation

# 8 Research Gaps

Globally, sustainability regulation is further on the rise. However, besides regulation that represents the legislative obligation of organizations to disclose sustainability information, aligning with societal discourse can be seen as a growing legitimacy obligation for companies and their sustainable actions and communication. Understanding and considering the role of the public and its related discourse not includes a "plurality of perspectives"[1]. It also relates to a shift of mindset that "creates a new basis of legitimacy and involves organizations in processes of active justification vis-à-vis society [and external stakeholders] rather than simply responding to the demands of powerful groups"[2].

Researchers stressed that the interpretation of sustainability in a business context is often vague and misleading. Banerjee (2003) emphasized that the problem of the broad definition of sustainability, such as the one in the Brundlandt Report, leads to controversies since human needs and wants are not considered well.[3] Though the Brundlandt Report mentioned social justice as a key focus of sustainability, Dobson (1998) stated a certain danger of mixing existent terms or even legitimizing a term against the other since "theories of sustainability sometimes stress the primacy of social justice, the position is often reversed"[4].

Since different interpretations exist, a structuring view that includes the different perspectives on the term sustainability and the relevance of materiality brings more clarity to the concepts. Lankoski (2016) presented four different focus areas and interpretations of the term sustainability in the business domain, namely first, sustainable development with its lens on its socioeconomic impact. Second, CSR

---

[1] Meynhardt (2019), p. 8.

[2] Palazzo and Scherer (2006), p. 72.

[3] see Banerjee (2003), p. 152; Kirkby et al. (1995).

[4] Dobson (1998), p. 242, as cited in Banerjee (2003), p. 152.



T. Heroth, *Corporate Sustainability Disclosure Matters*,
https://doi.org/10.1007/978-3-658-47653-3_8

concentrates on organizational actions. Third, a time-related focus on the long-term operability of companies, and fourth, a system-oriented view that considers the state of the system and the environment organizations are embedded in.[5]

The fourth point relates to the focus area of this dissertation since the environment of organizations, the value relevance of its stakeholders, and issues that drive organizational legitimacy are explored. This relates to the social dimension of sustainability. Ajmal et al. (2017) noticed that for explanations of sustainability in the business context, "the social dimension, however, has not yet been well explained"[6]. The authors emphasized growing societal pressure for organizations to address and explain negative social impacts.[7] This calls for clearer definitions of the social dimension. According to Bramley and Power (2019), an equitable society leaves no room for exclusion and prejudiced practices but allows full participation of individuals in economic and social activities.[8]

Recent research referring to stakeholder theory and its link to sustainability is manifold.[9] This link exists in a synergistic approach that aligns values with corporate stakeholders. "Stakeholder theory enlarges the scope to a broader societal embeddedness of organizations and its interdependencies with the societal environment"[10]. Studies that followed this approach mainly focused on sustainable business models[11] and sustainability accounting[12].

Dyllick and Hockerts (2002) interpreted corporate sustainability as "meeting the needs of a company's direct and indirect stakeholders (employees, clients, pressure groups, communities), without compromising its ability to meet the needs of future stakeholders as well"[13]. Already in the 1980s, a definition of corporate social

---

[5] Lankoski (2016), p. 849.

[6] Ajmal et al. (2018), p. 327.

[7] see Ajmal et al. (2018), p. 333.; Weingaertner and Moberg (2014).

[8] Bramley and Power (2009), as cited in Ajmal et al. (2018), p. 333.

[9] see Barrena Martínez et al. (2016); A. B. Carroll et al. 2013; Cornell and Shapiro (2021); Freudenreich et al. (2020); Hörisch et al. (2014); Hörisch et al. (2020); Schaltegger et al. (2019); Theodoulidis et al. (2017); Vitolla et al. (2019a); Vitolla et al. (2019b).

[10] Hörisch et al. (2014), p. 331.

[11] see Boons and Lüdeke-Freund (2013); Freudenreich et al. (2020); Lüdeke-Freund and Dembek (2017); Schaltegger et al. (2019); Schaltegger et al. (2022); Yang et al. (2017).

[12] see Christ et al. (2018); Hörisch et al. (2020).

[13] Dyllick and Hockerts (2002), p. 131.

responsibility by Jones (1980) described that "corporations have an obligation to constituent groups in society other than stockholders and beyond that prescribed by law or union contract, indicating that a stake may go beyond mere ownership"[14].

For assessing which stakeholders are of value relevance for companies, the analysis of stakeholder orientation helps to bring more light into the importance of the addressees of corporate disclosure. Studies that come closest to focus on the value relevance of stakeholders deal with stakeholders under the terms stakeholder engagement, stakeholder attention, stakeholder accountability, stakeholder dialogue, and stakeholder orientation.[15]

Focusing directly on the concept of stakeholder orientation, Bettinazzi and Zollo (2017) explored the orientation toward corporate stakeholders and its link to the long-term performance of mergers and acquisitions. The authors stated that a greater orientation towards stakeholders, such as employees, customers, suppliers, and local communities, affects performance characteristics of acquisitions, such as a positive effect of employees on the structural integration into the acquiring organization.[16] Stocker et al. (2020) explored the number of citations of stakeholders in corporate reporting by showing that the stakeholder employee is the most significant focus of companies in the energy sector. Followed by the community, shareholders, customers, suppliers, and the government, the authors identified different classification levels referring to the quality and extent of stakeholder engagement, "so the preparation of sustainability reports should seek a higher level of disclosure that enhances their effectiveness, going beyond GRI guidelines"[17].

However, to the best knowledge of the author of this dissertation, what needs to be added so far in research is an analysis of stakeholder orientation that includes various industries in a global sample aiming to detect its sustainable and financial value relevance. Empirical evidence that considers the importance of stakeholders such as society and the public is lacking. This dissertation tries to close this research gap by considering the shifts in societal discourse, focusing on the stakeholders involved.

To contribute to the welfare of individuals and to support society as a whole, organizations are asked to find the crucial link between their business activities and their alignment with societal needs.[18] Sustainability reporting, including stricter regula-

[14] Jones (1980), pp. 59-60.; Mitchell et al. (1997).

[15] see Bettinazzi and Zollo (2017); Bettinazzi and Feldman (2021); Crilly and Sloan (2012); Crilly and Sloan (2013); Harrison et al. (2010); Stocker et al. (2020); Zollo et al. (2009).

[16] Bettinazzi and Zollo (2017).

[17] Stocker et al. (2020), p. 2078.

[18] see Meynhardt et al. (2017); Meynhardt (2019).

tions, is, therefore, a driver that brings the concept of value and its understanding in a broader context since value creation is still inside but also motivated from outside the organization or, as Meynhardt and Bäro (2017) stated, "organizations cannot simply create 'value', as value has to be appreciated by the public"[19].

The question of when social sustainability would be guaranteed is answered by Littig and Grießler (2005), who stated that

> "if tasks within the general public and the related institutional courses of action fulfill a broadened set of human needs and are formed in a way that nature and its regenerative abilities are safeguarded over a drawn-out stretch of time and the regularizing cases of social equity, human pride, and cooperation are fulfilled"[20].

Therefore, the connection between sustainability and public value theory lies in the interpretation and perception of value. As sustainability enters the societal discourse about corporations, a value change in a broader sense is ignited, focusing on social issues and the societal impact of organizations. From a perspective of values and ethics, sustainability not only "honors the dignity of others (in this generation as well as generations to come), by treating the world as experienced as our collective inheritance"[21].

Public value helps to measure the points of reference that exist for organizations concerning their organizational legitimacy. According to Meynhardt (2019), "public value is more than reputation management, CSR or sustainability management. It is also not a zeitgeist-driven management paradigm. It helps us to recognise precisely the existing social functions of a business with all its paradoxes and potential contradictions"[22]. The societal discourse represents the radar that organizations need to incorporate on all business levels, including management, to detect, understand, and strategically act on the standards that are required for its societal legitimacy and acceptance.[23] However, since "these norms are not a given, but are in a constant state of flux, the corporate action process and the discourse on its appropriateness become decisive elements of responsible executive behavior"[24].

To also include the social issues that drive a corporate stakeholder relationship and create organizational legitimacy, the SDG framework as an approach towards double materiality is considered in the second empirical part of this dissertation,

---

[19] Meynhardt et al. (2017), p. 88.

[20] Littig and Griessler (2005), p. 11., as cited in Ajmal et al. (2018), p. 328.

[21] Bañon Gomis et al. (2011), p. 186.

[22] Meynhardt and Gomez (2019), p. 20.

[23] see Maak and Pless (2006).

[24] Meynhardt (2019), p. 408.

analyzing public value topic salience. According to a survey with a sample of around 200 global companies by GRI (2022), 83% speak of a general support of the SDGs, aiming to include it in sustainability reporting due to a commitment to the underlying values. 69% of companies differentiate concerning the relevance of SDGs, while 61% relate their corporate actions to the fulfillment of SDGs.[25]

Research papers that focused on SDGs in the context of business and sustainability reporting are still in their infancy while increasing recently. Rosati and Faria (2018) concentrated on company characteristics such as financials and board composition that are positively linked with a commitment towards SDG reporting.[26] Szennay et al. (2019) analyzed the alignment of SDGs with GRI indicators.[27] With the same focus on showing GRI and SDG linkages, Perello-Marin et al. (2022) focused on European car manufacturers and their disclosure of sustainability indicators and their mapping with the SDGs.[28] Pizzi et al. (2020) analyzed the orientation of companies toward the SDGs in corporate reporting with the help of an SDG reporting score. The authors stated that "the integration of the SDGs within the non-financial reports represents the first sign of being in favor of the diffusion of a more sustainable business model"[29] and a more excellent orientation towards and engagement with stakeholders. By concentrating on cultural factors by Hofstede, Pizzi et al. (2021) revealed that dimensions such as long-term orientation show a positive association with SDG reporting, while characteristics such as individualism are negatively related. Curtó-Pagès et al. (2021) explored SDG reporting of listed Spanish companies by showing significant progress of a greater orientation towards the SDGs in corporate reporting.[30] The SDGs in the Italian context of listed companies were analyzed by Izzo et al. (2020).[31]

All these publications focused on sustainability reporting and its link to the SDGs while contributing to the understanding of the SDGs as an approach towards double materiality. Studies that come closest to the empirical investigation of societal discourse, for example, refer to climate change discourse analysis, such as Fløttum and Gjerstad (2017), who explored the meaning of narratives in climate

[25] GRI and Support the Goals (2022).

[26] Rosati and Faria (2019).

[27] Szennay et al. (2019).

[28] Perello-Marin et al. (2022).

[29] Pizzi et al. (2021), p. 99.

[30] Curtó-Pagès et al. (2021).

[31] Izzo et al. (2020).

change discourse with the help of case studies in different geographical and political contexts.[32] Høvring et al. (2018) analyzed the role of discourse in corporate multi-stakeholder dialogues by referring to the role of power and positioning and the goal of stakeholders to raise their voice to shape discourses, while dissensus and conflict can be productive ways leading to social change. The authors highlighted the need for dialogue to create an understanding of underlying issues that are likely to exist in a corporate stakeholder relationship.[33]

Analyzing the SDGs with a view on societal discourse, Biermann et al. (2022) contributed empirical evidence that showed a discursive impact of SDGs on sustainable development in general, on its understanding and communication. The analysis of over 3000 studies related to SDGs revealed that the impact has limitations as a matter of a lack of transformational political measures and causality between reforms and observable outcomes. A greater interest of corporate actors is observable, though the results are not yet measurable concerning their effectiveness. The authors concluded that the first steps of a rising communication related to the SDGs and the Agenda 2030 can either be seen as a first step towards a more remarkable transformational change or a development that remains discursive, calling for research that empirically dives deeper into the effect of the SDGs on social and economic actors.[34]

How relevant are the SDGs as a framework for measuring public value and its influence on corporate sustainability and financial performance? Have social issues related to business organizations become a crucial component of societal discourse, being able to not only shape corporate agendas but also represent a relevant value criterion?

So far, the link between changes in societal discourse and its sustainable and financial value relevance of underlying issues that are linked to the single SDGs as a crucial driver of organizational legitimacy in the corporate context is not yet in circulation in the form of verifiable and empirical evidence. It is this "contextual sensitivity"[35] of companies that, according to Deetz and Simpson (2004), is essential, as "failure to attend to the 'otherness' around us limits our own perspective, produces incomplete and inadequate decision-relevant information, and does violence to those 'others' whose positions are often already institutionally and culturally marginalized"[36].

---

[32] Fløttum and Gjerstad (2017).

[33] Høvring et al. (2018).

[34] Biermann et al. (2022).

[35] Høvring et al. (2018), p. 641.

[36] Deetz and Simpson (2004), p. 157.

With a system perspective, some researchers call for a change of the underlying story and narrative of businesses, about stories tackling businesses specifically, the system organizations are embedded in, or about the values of the system as a whole"[37]. However, how do we know if the system, at least the understanding of organizational legitimacy, has not shifted yet? How do we know if social stakeholders that, according to public value theory, are, as part of a systemic view of organizations, not already a crucial decider for organizational legitimacy, including its sustainability and financial relevance? Though societal developments and the regulatory efforts of some governments might speak in favor of a shift that has already taken place, empirical evidence still needs to be explored and presented.

Therefore, this dissertation aims to close the two mentioned research gaps and provide evidence with the help of a content analytical approach and the analysis of disclosed information being part of corporate reports. The underlying research questions of the analysis are presented in the following chapter.

[37] see Lovins (2016).

# 9 Research Questions

As a conclusion of the research gaps that this dissertation focuses on, the following research question is put into the center of analysis:

*How value-relevant are stakeholder orientation and public value topic salience for sustainable and financial performance?*

Based on the research questions, this dissertation, with the help of empirical analysis, explores the following hypotheses to verify the value relevance of stakeholder orientation with a focus on sustainability performance:

> *H1a Social stakeholder orientation is positively associated with sustainability performance.*
> *H1b Financial stakeholder orientation is negatively associated with sustainability performance.*

Shifting the focus on financial performance and its association with stakeholder orientation leads to the following hypotheses:

> *H2a Social stakeholder orientation is negatively associated with the cost of capital.*
> *H2b Financial stakeholder orientation is positively associated with the cost of capital.*



T. Heroth, *Corporate Sustainability Disclosure Matters*,
https://doi.org/10.1007/978-3-658-47653-3_9

Concentrating on public value topic salience and its association with sustainability performance results in hypotheses H3 and H4.

*H3 Public value topic salience is positively associated with sustainability performance.*
*H4 Public value topic salience is negatively associated with the cost of capital.*

Taking corporate actions that are in favor of a more significant orientation towards its stakeholders and raising transparency by disclosing these actions and underlying issues in sustainability reports should influence the sustainability performance that results in favorable sustainability ratings and inversely influence the cost of capital since investors and banks intend to reward engagements and actions as well as transparency related to public value. The next chapter deals with the data sample of this dissertation.

# 10 Data Sample

The data sample consists of 90 companies that are part of six different industries, namely metals and mining companies in the materials sector, oil and gas companies in the energy sector, food companies in the consumer staples sector, textile and apparel companies in the consumer discretionary sector, software, and tech services companies of the technology sector and pharma companies of the pharmaceutical sector.

The industry selection refers to the Bloomberg Industry Classification Standard (BICS). Based on its market capitalization, the 15 biggest companies in the mentioned sectors are part of the data sample. The data sample includes sustainability reports that are published as ESG reports, CSR reports, corporate responsibility reports, impact reports, stakeholder impact reports, and corporate citizenship reports. Suppose none of the mentioned sustainability reports is published, in that case, the integrated report or annual report becomes part of the data sample.[1]

Figure 10.1 shows the distribution of the types of reports that are part of the data sample, with sustainability reports representing the majority (62.85%), followed by annual reports (27.37%), and integrated reports (9.78%)

[1] Although some companies publish summaries on ESG publications, they have not been included in my analysis due to a lack of comparability in terms of content and size.



T. Heroth, *Corporate Sustainability Disclosure Matters*,
https://doi.org/10.1007/978-3-658-47653-3_10

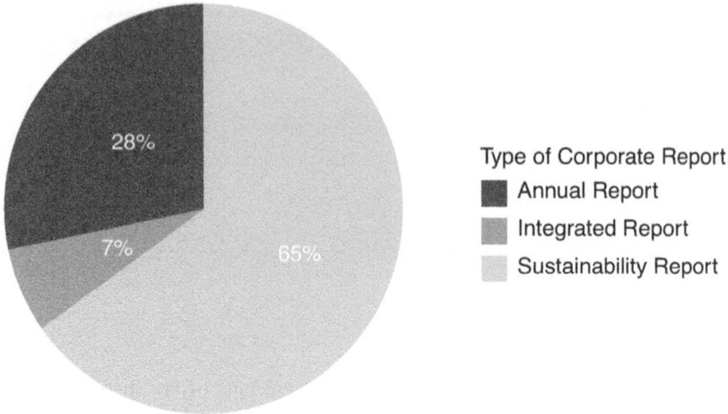


**Fig. 10.1** Distribution of Report Types in the Data Sample

The reports of the data sample cover the period of four years from 2018–2021. In total, the data sample contains 358 company reports collected from the company's webpages.[2]

As Fig. 10.2 illustrates, 139 reports originate from companies with headquarters in the United States, followed by Switzerland (32), the United Kingdom (28), France (28), China (20), India (16), and Germany (16). Canada follows with 12 reports, and Russia, Mexico, Japan, Brazil, and Australia with 8 reports. Saudi Arabia contributed 7 reports to the sample, while Norway, Italy, the British Virgin Islands, Ireland, and Denmark contributed 4 reports from companies headquartered in these respective countries.

The industry selection relates to the materiality of social topics defined by the SASB Standards of the Value Reporting Foundation of the IFRS Foundation. It provides a very conservative integration of industries and social topics into the data sample since most of those topics are included and are considered material from a single materiality perspective without requiring to be material in the sense of double materiality.

[2] In total, out of 90 companies and its published reports over four years, two reports are not part of the data sample since one company published its 2021 report in an online format without providing access to a downloadable PDF version and another company was not yet established in 2018.

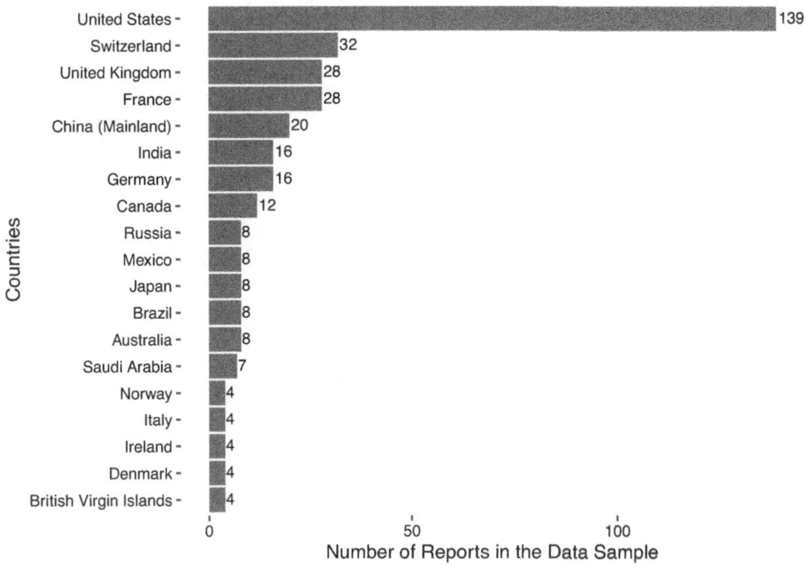


**Fig. 10.2** Geographic Distribution of Reports in the Data Sample According to Corporate Headquarters

The SASB materiality matrix of the Value Reporting Foundation shows a high relevance of social topics for the oil and gas and metals and mining sector with a focus on human rights and community relations. Social topics such as product quality and safety, customer welfare and selling practices, and product labeling play an essential role in the food industry. Product quality and safety issues are relevant for the textile and apparel sector, whereas software and tech services companies deal with customer privacy and data security.[3]

The following chapter dives into the empirical investigation, analyzing the constructs in focus with the help of the mentioned data sample.

[3] Value Reporting Foundation (SASB Standards) (2022).

# 11 Limitations of the Research in Focus

Every empirical research has its limitations. So does the one of this dissertation. Limitations refer to strategic management and communication reasons that can be traced back to the companies and the reports being part of the data sample, to the content of it, to technical reasons of the empirical approach, and cultural reasons.

Since we live in a time where developments in sustainability reporting are in flux, constructs such as stakeholder orientation and public value topics that are part of the documents in focus can be moving targets. One limitation of these circumstances is that sustainability reporting increasingly moves from a task of corporate communication departments to essential management responsibilities. This shift might have an influence on the way stakeholders are addressed and focused and the way topics are mentioned and prioritized. It takes time to develop from harnessing sustainability reporting as a means of creating corporate sustainable narratives and symbolic actions to create a strategic legitimacy to greater transparency of how management addresses societal stakeholders and positively reports public value. This might affect the results of the empirical analysis since the companies in the data sample are, to a very likely degree, in different phases of this process.

It is also evident that an analysis of corporate reports and, mainly, non-financial and sustainability reporting includes the limitation of a specific time lag. As Stocker (2020) concluded, it can be expected that dialogue and stakeholder involvement precede corporate disclosure on stakeholder orientation.[1] However, the real impact of disclosure regarding a direct relationship with sustainable corporate actions is hard to measure. It leaves room for the limitations of social washing or "SDG-Washing"[2], primarily when focusing on the quantification of mentions that relate to stakeholders or the number of SDGs.

---

[1] Stocker et al. (2020), p. 2073.

[2] Pizzi et al. (2021), p. 100.



T. Heroth, *Corporate Sustainability Disclosure Matters*,
https://doi.org/10.1007/978-3-658-47653-3_11

The data sample of this dissertation includes sustainability reports, annual reports, and integrated reports. This implicates different contents and audiences that might affect results when social stakeholders, for example, are less significant in annual reports than in sustainability reports. However, the selection of reports followed strict selection criteria, prioritizing sustainability reports over integrated and annual reports. Annual reports were only included in the data sample if no other comprehensive publication on sustainability existed. Another limitation that exists in that context is the way companies use different channels to communicate their sustainability efforts. Suppose a company publishes more content on its websites, social media, or ESG data channels. In that case, such content was not included in the analysis since the documents in focus were limited to reports officially published on a company's website in a report and PDF format.

In addition, since symbolic actions might still be part of some of the reports in the data sample, the chasm between what is communicated and what kind of action follows is outside the hands of the researcher. However, it still represents a limitation of empirical findings. With increasing transparency requirements, regulatory obligations, and pressure to maintain and build organizational legitimacy due to the external environment, it can be expected that more significant variance between disclosed information and action that follows will be discovered and antagonized.

For the data included in this dissertation, a limitation exists in content that might be influenced by recent standards published in sustainability reporting. Favoring the visibility of specific topics related to standards and guidelines pushed by standard setters might distort the picture of the constructs in the focus of the companies in the data sample and of this dissertation.

Technical limitations of an empirical exploration with the help of regression models often deal with the validity of the dependent, independent, and control variables. For the dependent variables that relate to sustainability performance, there are many more sustainability rating providers, such as MSCI, S&P, and ISS ESG, that already accounted for a total of over 600 providers in 2018,[3] and analyses show a great variety in terms of scope, methodologies and underlying weightings.[4] Therefore, possible deviating results with data from other rating providers represent a limitation. Nevertheless, sustainability ratings of two well-known and well-accepted rating providers, Refinitiv and Bloomberg, were included in the analysis to avoid subjectivity and to be able to verify the results.

For the independent variables related to stakeholder orientation, dictionaries with word lists for the different stakeholders result in the SOV scores. Although existing

[3] SustainAbility (2020).

[4] Berg et al. (2019).

academic publications and standard setters such as GRI were consulted to develop the dictionary, and synonyms as well as bigrams and trigrams were considered, textual information is never entitled to claim completeness. Therefore, if certain words for the stakeholders in focus are not part of the dictionary, an influence on the empirical results is within the bounds of possibility. For the SOV scores, only textual information was included in the score compilation. If stakeholder information was part of images or embedded in other data formats in the report, such information could not be further taken into account.

The independent variables that relate to public value topic salience depend on the mentions of SDGs in the underlying data sample. Since such a manual and numerical approach does not consider issues mentioned in other data sources, an influence on the final PVTS score is plausible.

Every regression model depends on the relevant control variables to be included in the analysis. With the control variables that are part of the empirical exploration of this dissertation, the most frequent variables were included, namely revenues, firm size, and performance metrics such as the EBITDA margin, to control for additional influencing factors that could distort the analysis from a financial and company perspective. In addition, sustainability-related control variables such as the Refinitiv ESG Policy Executive Compensation Score, which detects the relationship between management compensation and related ESG policies, and the UN Global Compact Membership score that shows if a company in the data sample is a UNGC member, were included. The variables are a control to see if there is an additional impact of other sustainability-related variables. Since these control variables raise no claims to completeness, it is essential to mention its possible limitations.

A cultural factor that might affect the empirical results is the influence of local reporting guidelines and frameworks. As an example, for the Chinese companies being part of the data sample, the Guidelines on Corporate Social Responsibility Reporting for Chinese Enterprises (CASS-CSR4.0) might be of higher relevance compared to standards issued by GRI. It could affect the way companies disclose sustainability content and how they address specific stakeholders in their reports.

Limitations might lead to distortions and biases of empirical results. However, to the best of the author's knowledge and possibilities, all counteracting measures to anticipate and avoid distortions caused by limitations were considered and realized. An empirical procedure, such as the content-analytical procedural model by Mayring (2014), helps to avoid distortions due to a lack of structure by providing proper and ordered steps to be part of the empirical investigation.[5]

---

[5] Mayring (2014).

# Empirical Analysis 12

This chapter analyzes stakeholder orientation and public value topic salience in corporate reporting with descriptive and inferential statistics to uncover the value relevance of both constructs in focus.

It contains the primary empirical investigations. The content-analytical procedural model by Mayring (2014) provides a transparent structure and a step-by-step approach.[1] The model helps to guide the reader through the empirical part of this dissertation and provides greater transparency of the steps that are part of the analysis.

## 12.1 Gaining Empirical Evidence with Content Analysis According to Mayring (2014)

Being pushed by the development of mass media in the thirties of the 20th century, analyzing data such as radio interviews and newspaper articles gained importance as an attempt "to discover public opinion"[2]. Content analysis is "the longest established method of text analysis among the set of empirical methods of social investigation"[3].

[1] Mayring (2014).
[2] Mayring (2014), p. 19.
[3] Titscher and Jenner (2000), p. 55.

**Supplementary Information** The online version contains supplementary material available at https://doi.org/10.1007/978-3-658-47653-3_12.



T. Heroth, *Corporate Sustainability Disclosure Matters*,
https://doi.org/10.1007/978-3-658-47653-3_12

Qualitative content analysis closely orientates toward research questions in focus and thereby, as a mixed method approach, differentiates itself from explanatory approaches such as grounded theory.[4]

## 12.2 Content-Analytical Procedural Model

The empirical research follows the model of Mayring (2014) to conduct a qualitative content analysis. Figure 12.1 shows the procedural model of a content analysis developed by Mayring.[5] Stemming from content analysis being developed to analyze large textual data and corpora out of the communication domain, close connections to linguistics and text analytics exist. Text analytics relates to Natural Language Processing (NLP) as a practical application of computational linguistics. Compared with data mining, which deals with the processing of structured data of databases, text analytics relates to the handling of unstructured or semi-structured data of text documents such as press articles, websites, blogs, and reports such as corporate disclosure.[6]

A deductive category application means that the categories related to stakeholders and public value in the form of the public value landscape[7] were previously classified by researchers. With the help of coding information from a raw data format into a reduced text format that "reduces text to a unit-by-variable format"[8], quantitative information can be gathered, and hypotheses can be tested based on qualitative text information.

With the procedural model, research becomes more comprehensible, reproducible, and testable for other researchers.[9] Since the underlying empirical research contains two empirical approaches, the procedural steps of the content analysis cover the components that are related to both explorations, the analysis of stakeholder orientation and the detection of public value topic salience. Corporate reports, being part of the data sample, are the focus material for both approaches. Therefore, the analysis of the document's origins, the formal characteristics of the material, and the determination of analyzing techniques as an overview of existing techniques are part of the initial phase of analysis. As shown in Fig. 12.1, the initial phase is

[4] Mayring (2014); Mayring (2019).

[5] Mayring (2014), p. 54.

[6] Moreno and Redondo (2016).

[7] Meynhardt (2009).

[8] Ryan and Bernard (2000), p. 780.

[9] Mayring (2014), p. 53.

part of the following section, followed by the categorical application and definition of analytical units for the subject's stakeholder orientation and public value topic salience.[10]

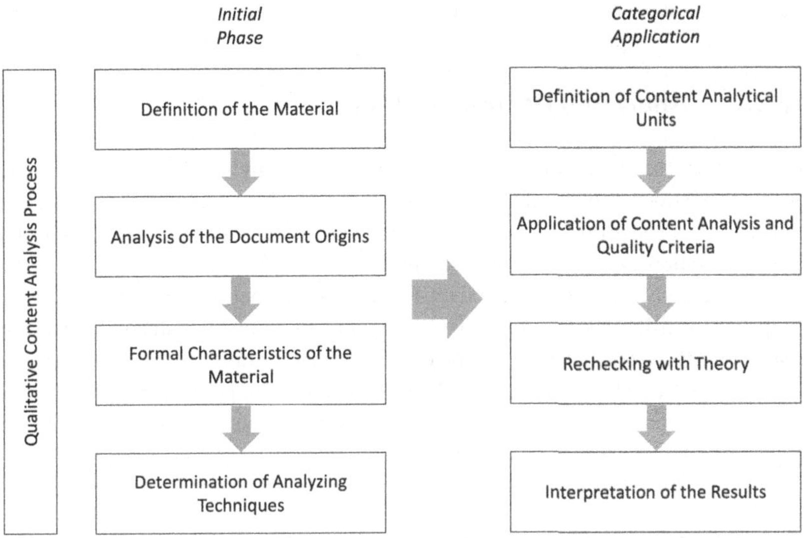


**Fig. 12.1** General Content-Analytical Procedural Model (Adapted From Mayring 2014)

### 12.2.1 Defining Sustainability Reports as Material in Focus

The general procedure starts with the selection of the material, which in the underlying research refers to documents published by the companies in the data sample, such as ESG reports, CSR reports, corporate responsibility reports, impact reports, stakeholder impact reports, and corporate citizenship reports, integrated reports or annual reports.

According to Mayring (2014), essential criteria for selecting the material in focus are, among others, the sample's representativeness, following a clear definition and selection logic.[11] The selection of documents is structured in a way that ESG, CSR, and sustainability reports are prioritized over other forms of corporate reports, such

[10] The structure of the step-by-step model by Mayring (2014) is attached to Appendix D in the Electronic Supplementary Material (ESM).

[11] Mayring (2014).

as annual reports, since non-financial information and the topics of stakeholder orientation and public value are the focus of this research and more likely to be part of sustainability disclosure. If no sustainability report of the company in focus exists, other report formats, such as impact reports, integrated reports, or annual reports, are added to the data sample.

### 12.2.2 Analyzing Document Origins

The documents in focus were retrieved from the company's web pages. Taking the origin of the documents in the data sample into account, the reports are published by the companies, primarily by departments focusing on communication, ESG, and investor relations, or by the management of a company and thereby represent the managerial and corporate perspective. There is no distinction between documents published by different departments or the document's author, which could affect the content disclosed and the analysis outcomes. The dissertation covers these restraints in the limitation section. However, all documents must be published on the official corporate webpage. The target group of the documents are external stakeholders, differing from investors to customers and the public.

### 12.2.3 Formal Characteristics and Quality Criteria of Sustainability Reports

Qualitative content analysis focuses on text categories, which means that it can only process the information that defined categories cover. Therefore, qualitative content analysis cannot include holistic expressions.[12] Focusing on the underlying text in the documents, first stored in a PDF format, a pdf-to-text transformer called *Pdftotext* allowed the transformation of the files from a PDF into a TXT format. If a sustainability report is only presented in an online version on a corporate webpage, the document was not included in the data sample.[13] The analysis includes the whole text of the reports and does not distinguish between certain sections. It is essential to highlight that the final document that is the subject of the content analysis can not contain pictures or figures since the text transcription cannot transfer these document

[12] Mayring (2019).

[13] Only one company of the data sample published its 2021 report in an online format without providing access to a downloadable PDF version.

formats into text. Due to that limitation, some information might not be considered for the final analysis.

### 12.2.4 Rule-based and Deductive versus Statistical and Inductive Analyzing Techniques

A rule-based or dictionary approach contains the definition of a list of keywords that are coded in a dictionary format to identify keyword groupings that, for example, can relate to topics, attributes, or entities. The underlying text corpus is analyzed with the help of mapping algorithms to detect words in a wordlist.[14] Wordlists are manually developed to analyze the construct in focus. Rule-based approaches suit research designs with a controllable sample size.

Analyzing the mention frequency of selected terms in a dictionary with an automated, rule-based approach to generate insights is a well-accepted approach in research. The benefits of automated text analytics compared to manual text analysis are a more objective exploration than manual content analysis since text information is automatically extracted and repeatable without manual sampling or tagging.[15] Besides, it allows the generation of detailed and novel insights based on a greater amount of complex and time-consuming data for manual extraction and analysis.[16]

Examples of similar approaches that focus on the frequencies of categories and words are Audi et al. (2016), who analyzed the role of trust as an aspect of corporate culture in annual reports.[17] Breuer et al. (2020) analyzed the Management Discussion and Analysis (MD&A) section of annual reports to explore the role of management rhetoric and trust-related words and their link to managerial opportunism and investor reactions.[18]

Statistical approaches in text analytics cover machine learning algorithms such as unsupervised methods with the aim "to learn 'hidden structures' in unlabeled data"[19]. Popular examples of unsupervised machine learning algorithms are topic modeling algorithms such as Latent-Dirichlet-Allocation (LDA) that explore topics in a corpus based on its Dirichlet distribution[20] or Bidirectional Encoder Represen-

---

[14] see Guo et al. (2016).

[15] Milne and Adler (1999).

[16] see Guo et al. (2016), Kwartler (2017).

[17] Audi et al. (2016).

[18] Breuer et al. (2020).

[19] Guo et al. (2016), p. 5.

[20] Blei et al. (2003).

tations from Transformers (BERT) and Large Language Models (LLMs)[21] that use a pre-trained language model for text classification.[22] Additional approaches include deep learning techniques, neural networks, support vector machines, or the Naïve Bayesian algorithm.[23] A machine-learning approach is advantageous if a large text data sample is subject to analysis, such as a Big Data sample. However, machine-learning approaches depend greatly on the pre-training of a training data set, which not only needs specialists for the pre-coding but also a reliable amount of training data.[24]

Text analytic studies and content analysis can follow a path of either deductive or inductive inference. A deductive inference bases its empirical approach with wordlists and dictionaries on existing hypotheses and theories, whereas an inductive approach bases its findings and theories on the generated data.[25]

In sum, a rule-based and deductive approach is preferred for this dissertation due to their replicability and transparency in generating and using categories based on existing research and their straightforward approach to generating insights of data samples that contain a precise number of company reports.[26] Therefore, the following section dives deeper into the analysis of stakeholder orientation with the application of a rule-based and deductive approach.

## 12.3 Application of Content Analysis I—Analyzing Stakeholder Orientation

This chapter deals with the first empirical analysis, exploring stakeholder orientation and its relevance for sustainability and financial value. Following Mayring's (2014) procedural model of content analysis, the first step is to define content analytical units that are the result of deductive reasoning based on existing research.

[21] Wei et al. (2022).

[22] Devlin et al. (2019).

[23] see Guo et al. (2016).

[24] Henry and Leone (2016), p. 159.

[25] Anandarajan et al. (2018), p. 16.

[26] Bannier et al. (2019), p. 85; cf. Hermes (2021).

### 12.3.1 Deductive Categorization—Stakeholders as Content Analytical Units

The selection of stakeholders is based on the most frequently mentioned stakeholders in the ESG reports that represent the content analytical units. The definition of words in focus and its classification is theory-based. Publications such as Stocker (2020) support the deductive approach to identify the most relevant stakeholders in the corporate context.[27] These theoretical findings are further backed by insights from standard setters such as GRI (2021) that define "common categories of stakeholders for organizations are business partners, civil society organizations, consumers, customers, employees, and other workers, governments, local communities, non-governmental organizations, shareholders and other investors, suppliers, trade unions, and vulnerable groups"[28]. The stakeholder categorization according to GRI relates to the GRI General Disclosure 2–29 guideline that highlights an approach to stakeholder engagement.[29]

Results have been cross-checked with the most frequently addressed stakeholders of a sample of sustainability reports. The final categorization of stakeholders includes the stakeholder association, bank, community, customer, employee, government, investor, media, NGO, shareholder, society, supplier, and university. To grasp words closely related to the mentioned stakeholders, synonyms and words most often mentioned as a substitute for a word category are part of the analysis. The selection of synonyms to compile the dictionary draws on synonym dictionaries using the remote package of NLTK WordNet,[30] being manually verified by checking the most frequently used words in the ESG reports of the data sample for the different stakeholders.

Stringently following coding rules in analyzing stakeholder orientation is no effort since the code is applied to all documents in the same form. Before generating the final stakeholder orientation scores, the SOVs of the different stakeholders, preprocessing steps include counting the total number of words in the document, generating word tokens, and transforming the text into a lowercase format. The content analytical and coding units are words, bigrams, or trigrams in the mentioned data sample. Plural forms of the different words are included in the dictionary.

[27] Stocker et al. (2020).

[28] GRI (2023), p. 47.

[29] GRI (2023).

[30] v.3.2.5.; Bird et al. (2009).

Table 12.1 shows the dictionary for generating the SOV of the mentioned stakeholders in the data sample that represents the calculation basis for creating the stakeholder SOV score for analyzing the stakeholder orientation of the companies in my data sample.

**Table 12.1** Stakeholder Dictionary With Main Stakeholders and its Related Terms

| Stakeholder_Dictionary |
|---|
| association |
| ['association'; 'trade association'; 'federation'; 'agency'; 'union'] |
| bank |
| ['banking partner'; banking institution'; 'creditor'; 'bond issuer'; 'depository'; 'financial service provider'] |
| community |
| ['neighborhood'; 'locality'; 'resident'; 'commonality'; 'commune'; 'citizen'] |
| customer |
| ['client'; 'buyer'; 'consumer'; 'purchaser'; 'user'; 'clientele'; 'merchant'; 'shopper'; 'subscriber'] |
| employee |
| ['staff'; 'employed person';'employed people'; 'jobholder'; 'colleague'; 'member of staff'; 'co-worker'; 'company member'; 'workforce'] |
| government |
| ['regulator'; 'administration'; 'authority'; 'council'; 'ministry'; 'jurisdiction'; 'regime'] |
| investor |
| ['financier'; 'creditor'; 'lender'; 'backer'; 'capitalist'; 'private equity'; 'business angel'; 'investment firm'] |
| media |
| ['press'; 'newspaper'; 'journalist'; 'news'; 'publishing house'; 'magazine'; 'columnist'; 'correspondent'; 'reporter'; 'commentator'] |
| ngo |
| ['non-governmental organization'; 'pressure group'; 'civil society organization'] |
| shareholder |
| ['equity owner'; 'share owner'; 'company owner'; 'stockholder'; 'participator'] |
| society |
| ['societal interest group'; 'opinion group'; 'public'; 'civil society'] |
| supplier |
| ['distributor'; 'provisioner'; 'purveyor'; 'dealer'; 'merchant'; 'trader'; 'tradesman'; 'retailer'; 'seller'; 'vendor'; 'wholesaler';'sub-contractor'] |
| university |
| ['academia'; 'academics'; 'academic body'; 'science'; 'research institution'] |

### 12.3.2 Quality Criteria

The dictionary is imported into a Python script that enables the counting of the mentioned frequency of the words of the dictionary and the total words of the single reports to be able to compile the SOV scores. The dictionary is defined and matched with the existing words in the documents. For scaling the analysis results, the SOV scores are multiplied by 100.[31] To avoid biases that are linked to word frequency and as a quality criterion, meaning that sustainability reports with greater length show higher results of words detected based on the dictionary, the empirical analysis focuses on the share of voice that puts mention frequency in context to the length of the document based on its total words. Dividing the frequency with the total number of words in the document leads to normalizing the frequency value.[32]

The following equation shows the calculation method for generating the SOV score.

$$SOV_{i,t} = \frac{\text{wordcount stakeholders}_{i,t}}{\text{total words}_{i,t}} X100$$

For efficiency gains, while running the analysis, the data sample is structured according to the different sectors. After exporting the SOV scores into a separate Microsoft Excel file, the results of the different sectors were transformed into one file that includes all the relevant variables for the analysis.

### 12.3.3 Descriptive Statistics: Stakeholder Orientation in Corporate Reporting

The first descriptive statistics part of the dissertation aims to answer the question of which stakeholders dominate as an addressee in the corporate reports of the data sample. It further explores yearly shifts of stakeholder orientation with the help of a cumulative view on the development of the SOV between 2018 and 2021, trying to detect certain momentum of stakeholders in corporate communication. Finally, the section gives a deeper insight into the sectoral differences in the representation of stakeholders. Figure 12.2 shows the total distribution for the SOV of the different stakeholders in the data sample. The figure shows clearly that a varying stakeholder orientation exists. The orientation toward employees as a stakeholder in corporate reporting dominates in the underlying data sample. This is backed by research such as Stocker et al. (2020), who identified employees as the most cited stakeholders

[31] see Breuer et al. (2020).

[32] Audi et al. (2016).

in their underlying sample.[33] Stakeholders that follow in terms of their SOV in the sample are community, customer, and supplier, with a two-digit share in the sample. The stakeholders, government, and society, are above a five percent threshold, while associations, universities, shareholders, investors, media, banks, and NGOs range from one to four percent share. In sum, the stakeholder employee dominates the SOV in the data sample, while other social stakeholders, such as community and business-related stakeholders, such as customers and suppliers, follow.

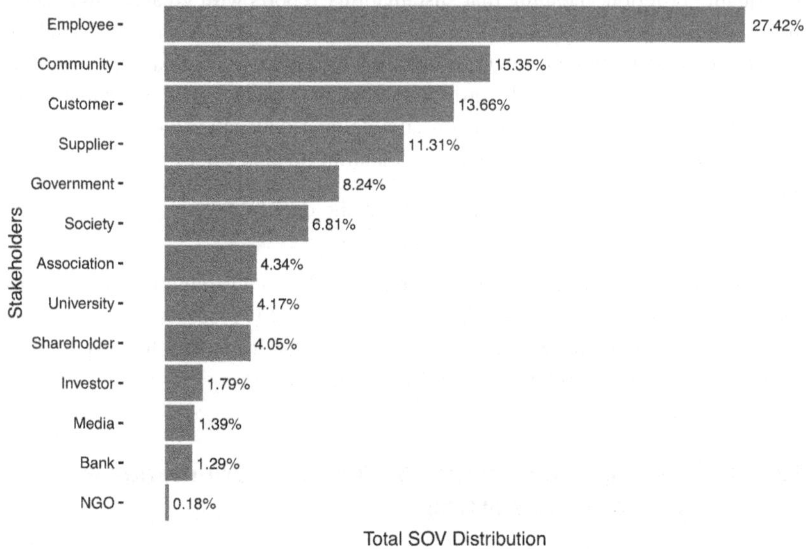


**Fig. 12.2** Distribution of Stakeholders in the Data Sample

Basing the categorization into social and financial stakeholders on the findings in the theoretical part of this dissertation, financial stakeholders including shareholder, investor, and bank account for a 7.13% share, while social stakeholders with the stakeholders employee, community, society, and NGO have a share of 49.76% in the data sample. This shows a great representation of this stakeholder category in the underlying corporate sustainability reports, mainly driven by the stakeholders

[33] Stocker et al. (2020), p. 2077.

employee and community, followed by society, with a neglectable representation of the SOV for NGO. The remaining part of 43.11% is covered by the stakeholders customer, supplier, government, association, university, and media that play a role in corporate reporting but are not part of the classification of social and financial stakeholders.

This raises the question of the sectoral picture of stakeholder orientation over the six different sectors of the data sample. As Fig. 12.3 shows, stakeholder orientation is sector-dependent, but the general distribution of the total sample can be confirmed with the sectoral view.

The stakeholder employee dominates almost all sectors despite a 1% difference to the stakeholder customer in the food sector. The highest SOV for the stakeholder employee is in the pharma (32%) and tech sector (30%). Community as a stakeholder plays a significant role in sectors such as metals and mining (23%) and in the food industry (20%), compared to a lower SOV in the pharma (8%) and textile sector (10%). Besides its highest SOV in the food sector, with the same share of 21%, the stakeholder customer is well represented in the tech sector (21%), followed by the textile industry (14%).

The highest SOV for supplier can be found in the textile (19%), food (14%), tech (11%), and pharma (10%) industries, while the stakeholder government is in focus of sectors such as oil and gas (13%) and the metals and mining industry (12%). The SOV for the stakeholder society is well represented in the pharma (10%), the oil and gas (8%), and metals and mining industry (8%). All other stakeholders only have a one-digit percentage in the share of the total distribution of SOV. Financial stakeholders with lower SOV are mainly represented by the stakeholder shareholder with a 6% share in the oil and gas and textile industry.[34]

In sum, apart from the food sector in which the stakeholder customer has the highest SOV, it can be highlighted that the stakeholder employee dominates the sectors regarding its SOV and representation in the data sample. Besides further social and financial stakeholders, the stakeholders customer, supplier, and government are a key focus group in the data sample.

[34] An illustration of the mention frequency results on a company level for the textile industry is added to the Appendix C of this dissertation in the Electronic Supplementary Material (ESM).

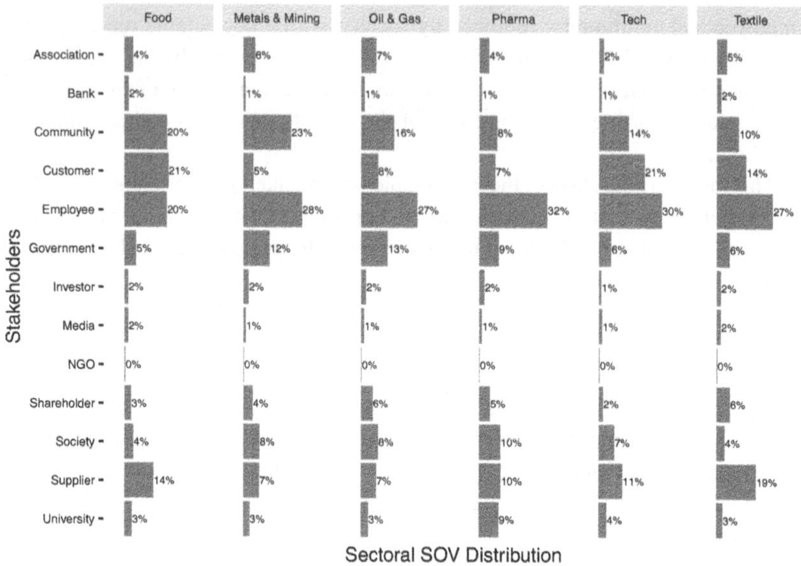


**Fig. 12.3** Sectoral Distribution of Stakeholder Orientation

It must be acknowledged that the orientation towards stakeholders in corporate reporting varies over time. As Fig. 12.4 shows, cumulative shifts between 2018 and 2021 reveal that only the stakeholders NGO, supplier, and employee gained greater attention in the form of a growing SOV. Due to the low total SOV of the stakeholder NGO in the data sample, the increase of this stakeholder group is driven by a small number of companies, mainly by an initiative of a company in the tech industry, and can be neglected for further analysis.

In sum, the SOV growth of the stakeholder employee confirms its leading role as a stakeholder in corporate reporting. In addition, business-related stakeholders such as the stakeholder supplier gained momentum in recent years, at least regarding the sectors in focus of the underlying data sample.

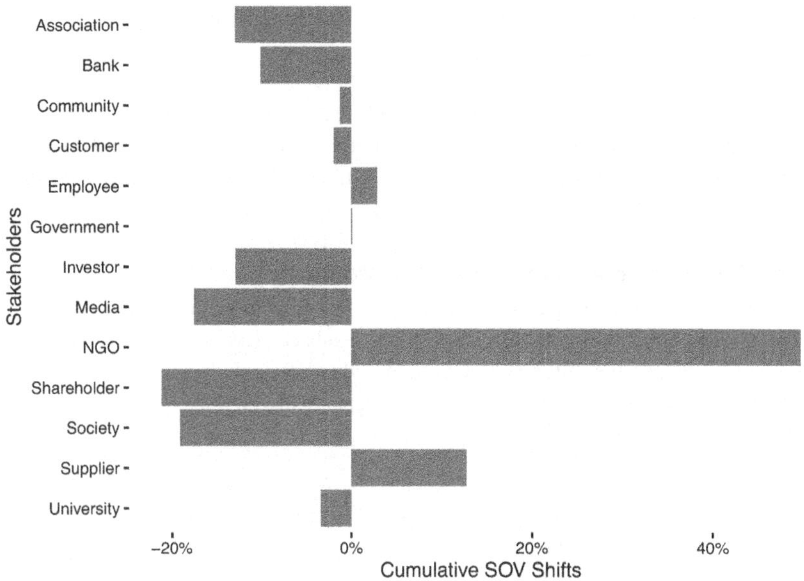


**Fig. 12.4** Cumulative Stakeholder Shifts in the Data Sample

Guiding the research closer to generating inferential insights, the classification between social and financial stakeholders helps to gain knowledge about the sustainable and financial value relevance of stakeholder orientation.

The sectoral and categorized view as in Fig. 12.5 shows an even distribution of social stakeholders among the sectors in focus, all reaching at least 41% of the total sector distribution. The highest SOV distribution of the stakeholders employee, community, society, and NGO is in metals and mining (59%), followed by oil and gas (51%), and the tech sector (51%). The remaining sectors, food, and textile, reach a distribution of 45% and 41%.

Financial stakeholders, in comparison, have lower total SOV as shown in Fig. 12.6 and are most prominent in the textile industry (10%), followed by a one-digit share in oil and gas (9%), pharma (8%), metals and mining (7%), and the food sector (6%). The tech sector (4%) is least oriented towards financial stakeholders regarding SOV in the underlying sample.

### 12.3.4 Inferential Statistics—The Value Relevance of Stakeholder Orientation

The first inferential statistics part aims to explore the sustainable and financial value relevance of stakeholder orientation, differentiating between social and financial stakeholders. The hypotheses are, therefore, embedded in the following regression models and enriched by additional variables.

#### 12.3.4.1 Variables and Regression Models

The analysis of stakeholder orientation as a relevant influencing factor on sustainability and financial performance considers a subset of variables. The following Table 12.2 shows the summary of variables that are included for the inferential statistics part of my analysis, clustered into dependent variables, independent variables, and control variables.[35]

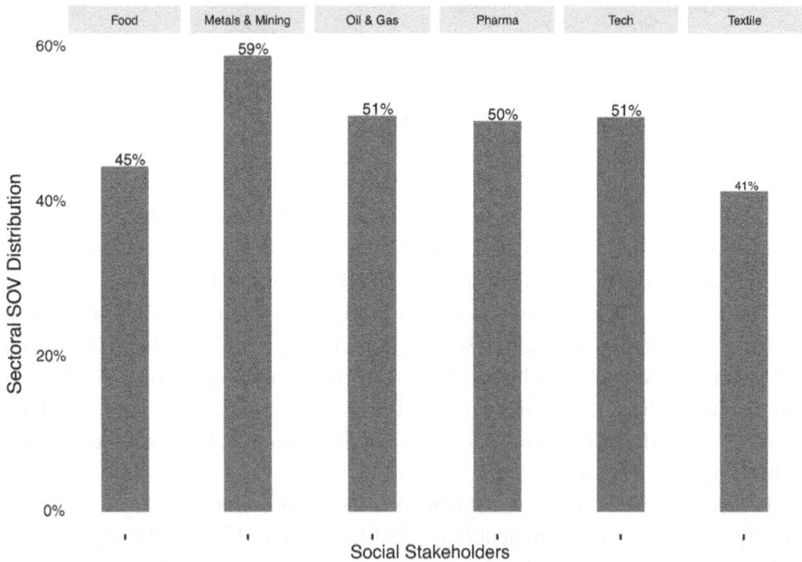


**Fig. 12.5** Social Stakeholder Distribution According to the six Different Sectors

[35] All single inferential statistic regressions are part of Appendix A of this dissertation in the Electronic Supplementary Material (ESM).

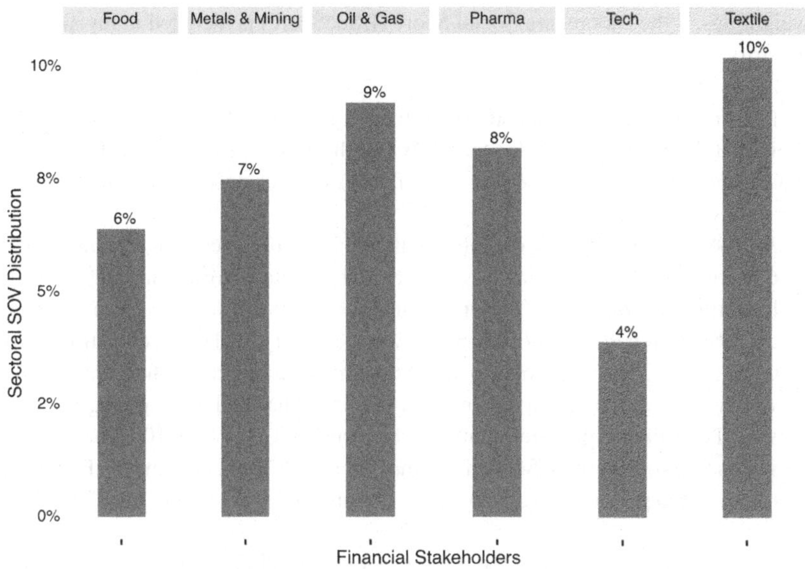


**Fig. 12.6** Financial Stakeholder Distribution According to the six Different Sectors

**Dependent Variables** The dependent variable in the underlying regression models that explore sustainability performance is the ESG score provided by the third-party data provider Refinitiv (RESGScore). Since ESG ratings came under scrutiny due to dispersing results in comparison with each other, mainly due to different scopes and

**Table 12.2** List of Variables, Domains and Scores Applied in the Regression Model for Stakeholder Orientation

| Variables | Domain | Score |
|---|---|---|
| Dependant Variables | Sustainability Performance | RESGScore |
| | Financial Performance | WACC |
| | | WACCCostofEquity |
| | | WACCCostofDebt |
| Independent Variables | Stakeholder Orientation | SOV |
| Control Variables | ESG-Related | BSocScore |
| | | PolicyExecutiveCompensationScore |
| | Market-Related | lnRevenue |
| | Performance-Related | EBITDAMargin |

weightings,[36] the Bloomberg Social Score (BSocScore) is included in my analysis to verify the analytical approach.

For the analysis of financial performance, the cost of capital, including the Weighted Average Cost of Capital (WACC), the cost of equity (WACCCostofEquity), and the cost of debt (WACCCostofDebt) serve as dependent variables.

*RESGScore*: The RESGScore refers to the Refinitiv ESG Score and entails around 630 company-level ESG metrics that are divided into ten different categories, illustrating "a company's ESG performance, commitment and effectiveness based on publicly-reported information"[37]. Every category, under consideration of its category weights, contributes to the ESG score. In comparison, the S and E pillar weights differ according to the industry in focus. The G pillar weighting does not vary. The pillar weights are summarized into an ESG score of 0–100. The D Score stands for poor relative ESG performance for ratings between 0 and 25. Excellent relative ESG performance, the A Score, is equal to scores ranging from 75 to 100.

*WACC*: The variable Weighted Average Cost of Capital (WACC) is a metric for measuring a company's cost of capital as provided by Refinitiv. It includes stocks (equity and preferred) and debt; each capital source is weighted proportionately and expressed as a percentage.

*WACCCostofEquity*: This variable refers to the weighted equity costs of the WACC. It refers to the risk premium that a company pays to its equity investors. The calculation of the cost of equity results from adding the multiplied equity risk premium of the market with a company's beta factor of the stock to the inflation-adjusted risk-free rate, expressed as a percentage.

*WACCCostofDebt*: This variable refers to the weighted debt costs of the WACC. It includes the marginal cost for the company to issue new debt and includes the weighted cost of short-term and long-term debt, expressed as a percentage.

**Independent Variables** The independent variables in the regression model for analyzing stakeholder orientation contain the following metrics.

[36] see Berg et al. (2019), Kotsantonis and Serafeim (2019).

[37] Refinitiv (2022), p. 6.

*SOV_society*: SOV_society is the share of voice for the stakeholder society, compounded by its mention frequency divided by the total number of words in the respective report and normalized by multiplying with 100, expressed in decimal form.

*SOV_bank*: SOV_bank is the share of voice for the stakeholder bank, compounded by its mention frequency divided by the total number of words in the respective report and normalized by multiplying with 100, expressed in decimal form.

*SOV_community*: SOV_community is the share of voice for the stakeholder community, compounded by its mention frequency divided by the total number of words in the respective report and normalized by multiplying with 100, expressed in decimal form.

*SOV_employee*: SOV_employee is the share of voice for the stakeholder employee, compounded by its mention frequency divided by the total number of words in the respective report and normalized by multiplying with 100, expressed in decimal form.

*SOV_shareholder*: SOV_employee is the share of voice for the stakeholder shareholder, compounded by its mention frequency divided by the total number of words in the respective report and normalized by multiplying with 100, expressed in decimal form.

**Control Variables** Market-related variables such as revenues and serve as controlling variables in the regression model. LnRevenue is considered with their natural logarithm to control for too much influence on the regression results of companies with large revenues. The additional control variables BSocScore to verify the validity of evaluating ESG performance concerning the social pillar and the Policy Executive Compensation ESG Performance Score provided by Refinitiv act as a control for extra financial performance related to corporate management. As a financial performance metric, the EBITDA margin of the companies in the sample is included in the models.

*BSocScore*: The variable BSocScore refers to the Bloomberg Social Pillar Score. Bloomberg's ESG Pillar Scores are a weighted mean of issue scores ranked according to the industries in focus. Bloomberg states that "the ranking of E and S pillars were assessed based on Bloomberg Intelligence analysis of the relative exposure to E and S issues identified in the industry's materiality framework"[38]. The social score goes from 0 to 10, while higher numbers indicate a better social performance rated by Bloomberg.

*PolicyExecutiveCompensationScore*: This score represents an extra-financial performance component provided by Refinitiv and contains a management compensation factor related to ESG and sustainability in a range between 0 and 100.

*lnRevenue*: lnRevenue is the natural logarithm of the yearly revenues in a fiscal period of the specific company in focus, expressed in decimal form.
*EBITDAMargin*: The variable EBITDAMargin is the result of the yearly EBITDA in a fiscal period of the specific company in focus divided by its total revenue for the same period, expressed in percentage.

#### 12.3.4.2 Stakeholder Orientation and Sustainability Performance

To analyze the influence of corporate stakeholder orientation on sustainability performance with the help of ESG ratings, a linear regression of the panel data is conducted. The panel data includes company identification codes that support the analysis of 90 companies in total over the defined time frame. The time-related panel settings refer to the publication dates of the sustainability reports. Random effect and fixed effect model types are considered in the analysis.[39] The Hausman Specification Test at the 5% statistical significance level further specifies model selection. Robust estimates as standard error type are included in the analysis to control for heteroskedasticity-consistent standard errors.[40]

For every regression model, two tests help to check for normal distribution of residuals and test for multicollinearity of the included variables. First, the Jarque-Bera test helps to find out whether the characteristics of skewness and the kurtosis of the sample data suit a normal distribution. A p-value exceeding the implied significance level of 5% shows no violation of the assumption of normal distribution, meaning that residuals are normally distributed.[41] Second, multicollinearity is tested

---

[38] Bloomberg (n.d.), p. 15.

[39] see example of test statistics in Appendix B in the Electronic Supplementary Material (ESM).

[40] Arellano (1993).

[41] Sajwan and Priya (2018).

for the included variables with the help of uncentered variance inflation factors (VIF).[42]

Since heteroscedasticity is a problem in many cross-sectional regressions,[43] the variance of the errors across the given observations can be controlled for by considering heteroskedasticity-robust standard errors that are included in the regression models.[44]

Considering the mentioned statistical tests, the first hypothesis refers to the sustainable value relevance of social stakeholders. It tests whether the mentioning of social stakeholders has a positive effect on the dependent variable, the RESGScore.

*H1a Social stakeholder orientation is positively associated with sustainability performance.*

To test the first hypothesis, H1a, a panel regression tests all social stakeholders on its significance and correlation with the Refinitiv ESG Score (RESGScore) by controlling for the Bloomberg Social Score (BSocScore) and lnRevenue. The SOVsocialstakeholders refers to all the share of voices of the social stakeholders included in the model, namely society, community, employee, NGO, association, and media.

To guarantee comparability between the included ratings and to verify the regression model with an additional sustainability metric, the BSocScore is considered in the analysis. In addition, the control variable lnRevenue controls for the influence of revenue size.

The equation of the regression model testing H1a looks as follows.

$$Model(1)$$

$$RESGScore_{i,t} = \alpha_i + \beta_1[SOVsocialstakeholders]_{i,t}$$

$$\beta_2[BSocScore]_{i,t} + \beta_3[lnRevenue]_{i,t} + u_{i,t}$$

Table 12.3 shows that the stakeholder society is positively correlated with the RESGScore. With a modest significance level below 10% and solid goodness of fit ($R^2$) of 22.7%, hypothesis H1a can be verified for the stakeholder society. Both control variables, the BSocScore and lnRevenue are significant, showing a positive correlation with the RESGScore.[45]

---

[42] see Backhaus et al. (2018); Belsley (1991).

[43] Long and Ervin (2000).

[44] see Kezdi (2004), Stock and Watson (2008).

[45] In the regressions where the sample size is reduced to 268 observations, the included ratings were not available for the whole sample period but only for 2018–2020.

**Table 12.3** The Correlation of the Stakeholder Society With the RESGScore

| Model (re) | (1) |
|---|---|
| Variables | RESGScore |
| SOV_society | 8.472* |
| | (5.016) |
| BSocScore | 1.124** |
| | (0.437) |
| lnRevenue | 3.474*** |
| | (1.184) |
| Constant | −14.49 |
| | (27.92) |
| Observations | 268 |
| R-squared (between) | 0.227 |
| Number of companycode | 90 |

*Note.* All models are estimated as robust standard errors (in parentheses). Model selection is based on a generalized Hausman Test, selecting between a GLS random effects (re) and a fixed effects model (fe). *,** and *** indicate statistical significance at the 10%, 5% and 1% levels, respectively.

*H1b Financial stakeholder orientation is negatively associated with sustainability performance.*

The second hypothesis aims to shed more light on the interlinks between the orientation towards financial stakeholders and its influence on sustainability performance. An additional panel regression tests all financial stakeholders on its significance and correlation with the Refinitiv ESG Score (RESGScore) by controlling for the Bloomberg Social Score (BSocScore) and lnRevenue. Financial stakeholders include the stakeholders investor, bank, and shareholder.

The equation of the regression model testing H1b looks as follows.

$$Model(2)$$

$$RESGScore_{i,t} = \alpha_i + \beta_1[SOVfinancialstakeholders]_{i,t} + \beta_2[BSocScore]_{i,t} + \beta_3[lnRevenue]_{i,t} + u_{i,t}$$

Table 12.4 reveals that that there is a significant, negative correlation between the mentioning of banks in the reports and the RESGScore of the individual companies. Being significantly below the 5% level and with a solid goodness of fit of 21.6% ($R^2$), hypothesis H1b can be verified for the stakeholder bank. Both control variables, the BSocScore and lnRevenue, are significantly and positively correlated with the RESGScore.[46]

**Table 12.4** The Correlation of the Stakeholder Bank with the RESGScore

| MODEL (re) | (2) |
|---|---|
| VARIABLES | RESGScore |
| SOV_bank | −50.33** |
| | (24.15) |
| BSocScore | 1.171*** |
| | (0.451) |
| lnRevenue | 3.429*** |
| | (1.181) |
| Constant | −12.17 |
| | (28.02) |
| Observations | 268 |
| R-squared (between) | 0.216 |
| Number of companycode | 90 |

*Note.* All models are estimated as robust standard errors (in parentheses). Model selection is based on a generalized Hausman Test, selecting between a GLS random effects (re) and a fixed effects model (fe). *,** and *** indicate statistical significance at the 10%, 5% and 1% levels, respectively.

##### 12.3.4.3 Stakeholder Orientation and Financial Performance

To explore whether financial performance in the form of a company's cost of capital is also significantly correlated with social and financial stakeholder orientation, H2a and H2b are the related hypotheses that are further explored based on the regression models.

*H2a Social stakeholder orientation is negatively associated with the cost of capital.* The exploration of the relationship between stakeholder orientation and financial performance in the form of the cost of capital is tested regarding hypothesis H2a, which states that social stakeholder orientation is negatively associated with the cost of capital. A negative association means that social stakeholder orientation has a positive effect on the cost of capital since companies can benefit from lower cost of capital, incurring lower costs to finance a business. Variables for the cost of capital include the Weighted Average Cost of Capital (WACC)[47], the Weighted Average

[46] Other control variables were excluded due to a non-significant relationship.

[47] Model 3,6,9.

Cost of Equity (WACCCostofEquity)[48], and the Weighted Average Cost of Debt (WACCCostofDebt)[49].

The regression models for analyzing H2a are shown as follows.[50] SOVsocialstakeholders include all the share of voices for the social stakeholders included in the model.

$$Model(3, 6, 9)$$

$$WACC_{i,t} = \alpha_i + \beta_1[SOVsocialstakeholders]_{i,t} +$$
$$\beta_2[PolicyExecutiveCompensationScore]_{i,t} +$$
$$\beta_3[EBITDAMargin]_{i,t} + \beta_4[lnRevenue]_{i,t} + u_{i,t}$$

$$Model(4, 10)$$

$$WACCCostofEquity_{i,t} = \alpha_i + \beta_1[SOVsocialstakeholders]_{i,t} +$$
$$\beta_2[PolicyExecutiveCompensationScore]_{i,t} +$$
$$\beta_3[EBITDAMargin]_{i,t} + \beta_4[lnRevenue]_{i,t} + u_{i,t}$$

$$Model(7)$$

$$WACCCostofEquity_{i,t} = \alpha_i + \beta_1[SOVsocialstakeholders]_{i,t} +$$
$$\beta_2[PolicyExecutiveCompensationScore]_{i,t} + \beta_3[lnRevenue]_{i,t} + u_{i,t}$$

[48] Model 4,7,10.

[49] Model 5,8,11.

[50] For models 5, 7, 8, and 11, the control variable EBITDAMargin has been excluded due to non-significance.

$$Model(5, 8, 11)$$

$$WACCCostofDebt_{i,t} = \alpha_i + \beta_1[SOVsocialstakeholders]_{i,t} + \beta_2[PolicyExecutiveCompensationScore]_{i,t} + \beta_3[lnRevenue]_{i,t} + u_{i,t}$$

Table 12.5 shows that the regression model referring to the social stakeholder society reveals the most robust results for a negative correlation with the WACCCostofDebt with high significance below the 1% level and a solid $R^2$ of 24.1%. With a significance below 5% for the WACC and below 10% for the WACCCostofEquity and an $R^2$ of 13.1% and 9.8%, respectively, the SOV_society is negatively correlated to all costs of capital, revealing a positive impact of social stakeholder orientation when addressing the stakeholder society.

**Table 12.5** The Correlation of the Stakeholder Society With the Cost of Capital

| MODEL (fe) | (3) | (4) | (5) |
|---|---|---|---|
| VARIABLES | WACC | WACCCostofEquity | WACCCostofDebt |
| SOV_society | −0.0686** | −0.0732* | −0.0739*** |
| | (0.0343) | (0.0405) | (0.0197) |
| PolicyExecutive | −0.0114** | −0.0117** | −0.00648* |
| CompensationScore | (0.00502) | (0.00500) | (0.00378) |
| EBITDAMargin | 0.0842*** | 0.0913** | |
| | (0.0317) | (0.0449) | |
| lnRevenue | 0.0290*** | 0.0271*** | 0.0299*** |
| | (0.00766) | (0.00939) | (0.00334) |
| Constant | −0.646*** | −0.590*** | −0.686*** |
| | (0.183) | (0.223) | (0.0813) |
| Observations | 358 | 358 | 358 |
| R-squared (within) | 0.131 | 0.098 | 0.241 |
| Number of companycode | 90 | 90 | 90 |

*Note.* All models are estimated as robust standard errors (in parentheses). Model selection is based on a generalized Hausman Test, selecting between a GLS random effects (re) and a fixed effects model (fe). *,** and *** indicate statistical significance at the 10%, 5% and 1% levels, respectively. Control variables that are not significant are excluded

In addition, the social stakeholder community shows significant results in relation with the cost of capital. As Table 12.6 reveals, the stakeholder community (SOV_community) is negatively correlated with the highest significance level below 1% and an $R^2$ of 20.7% for the WACCCostofDebt. For the WACCCostofEquity significance is below the 5% level, and for the WACC below 10%.

**Table 12.6** The Correlation of the Stakeholder Community With the Cost of Capital

| MODEL (fe) | (6) | (7) | (8) |
|---|---|---|---|
| VARIABLES | WACC | WACCCostofEquity | WACCCostofDebt |
| SOV_community | −0.0281* | −0.0368** | −0.0199*** |
| | (0.0148) | (0.0176) | (0.00698) |
| PolicyExecutive | −0.0116** | −0.0114** | −0.00670* |
| CompensationScore | (0.00499) | (0.00510) | (0.00359) |
| EBITDAMargin | 0.0737** | | |
| | (0.0357) | | |
| lnRevenue | 0.0306*** | 0.0371*** | 0.0301*** |
| | (0.00830) | (0.00964) | (0.00393) |
| Constant | −0.682*** | −0.804*** | −0.692*** |
| | (0.196) | (0.233) | (0.0951) |
| Observations | 358 | 358 | 358 |
| R-squared (within) | 0.127 | 0.081 | 0.207 |
| Number of companycode | 90 | 90 | 90 |

*Note.* All models are estimated as robust standard errors (in parentheses). Model selection is based on a generalized Hausman Test, selecting between a GLS random effects (re) and a fixed effects model (fe). *,** and *** indicate statistical significance at the 10%, 5% and 1% levels, respectively.

As a third stakeholder in the underlying classification that falls under the social category, the stakeholder employee shows significant results in the regression with capital costs. Table 12.7 indicates a significance below 5% for the WACCCostofDebt with the highest $R^2$ of 20.9%, and significance levels below 10% for the WACC and the WACCCostofEquity with $R^2$ of 13.1% and 9.8% respectively.

**Table 12.7** The Correlation of the Stakeholder Employee With the Cost of Capital

| MODEL (fe) | (9) | (10) | (11) |
|---|---|---|---|
| VARIABLES | WACC | WACCCostofEquity | WACCCostofDebt |
| SOV_employee | −0.0305* | −0.0340* | −0.0184** |
| | (0.0170) | (0.0184) | (0.00709) |
| PolicyExecutive | −0.0126** | −0.0130** | −0.00724** |
| CompensationScore | (0.00495) | (0.00500) | (0.00343) |
| EBITDAMargin | 0.0792** | 0.0858* | |
| | (0.0342) | (0.0479) | |
| lnRevenue | 0.0298*** | 0.0279*** | 0.0297*** |
| | (0.00797) | (0.00986) | (0.00405) |
| Constant | −0.657*** | −0.603** | −0.682*** |
| | (0.189) | (0.233) | (0.0980) |
| Observations | 358 | 358 | 358 |
| R-squared (within) | 0.131 | 0.098 | 0.209 |
| Number of companycode | 90 | 90 | 90 |

*Note.* All models are estimated as robust standard errors (in parentheses). Model selection is based on a generalized Hausman Test, selecting between a GLS random effects (re) and a fixed effects model (fe). *,** and *** indicate statistical significance at the 10%, 5% and 1% levels, respectively.

Consequently, hypothesis H2a can be verified for the social stakeholders, society, community, and employee. The SOV for society shows robust results and the highest correlation factors for all three capital costs, with the highest significance for the WACCCostofDebt. This implies a negative association and a beneficial impact of stakeholder orientation toward the stakeholder society on the capital costs related to debt.

For the WACC and the WACCCostofEquity, the highest correlation exists for the stakeholder society, while the stakeholder community reaches the most significant results for equity financing.

*H2b Financial stakeholder orientation is positively associated with the cost of capital.*

Comparing the effect of financial stakeholder orientation on the cost of capital and potential adverse cost effects that might exist when addressing stakeholders such as the stakeholders bank, investor, and shareholder is explored with the help of the following regression equations.

$$Model(12)$$

$$WACC_{i,t} = \alpha_i + \beta_1[SOV financialstakeholders]_{i,t} +$$
$$\beta_2[PolicyExecutiveCompensationScore]_{i,t} +$$
$$\beta_3[EBITDAMargin]_{i,t} + \beta_4[lnRevenue]_{i,t} + u_{i,t}$$

$$Model(13)$$

$$WACCCostofEquity_{i,t} = \alpha_i + \beta_1[SOV financialstakeholders]_{i,t} +$$
$$\beta_2[PolicyExecutiveCompensationScore]_{i,t} +$$
$$\beta_3[EBITDAMargin]_{i,t} + \beta_4[lnRevenue]_{i,t} + u_{i,t}$$

As Table 12.8 indicates, hypothesis H2b can be verified for the stakeholder shareholder, showing a positive correlation with the WACC (Model 12) and the WACCCostofEquity (Model 13). With higher correlation coefficients for the WACCCostofEquity, the significance level for both capital costs is below the 10% level

**Table 12.8** The Correlation of the Stakeholder Shareholder With the Cost of Capital

| MODEL (fe) | (12) | (13) |
|---|---|---|
| VARIABLES | WACC | WACCCostofEquity |
| SOV_shareholder | 0.105* | 0.125* |
| | (0.0579) | (0.0674) |
| PolicyExecutiveCompensationScore | −0.0120** | −0.0124** |
| | (0.00481) | (0.00477) |
| EBITDAMargin | 0.0790** | 0.0851* |
| | (0.0326) | (0.0465) |
| lnRevenue | 0.0321*** | 0.0308*** |
| | (0.00776) | (0.00983) |
| Constant | −0.728*** | −0.689*** |
| | (0.185) | (0.233) |
| Observations | 358 | 358 |
| R-squared (within) | 0.131 | 0.100 |
| Number of companycode | 90 | 90 |

*Note.* All models are estimated as robust standard errors (in parentheses). Model selection is based on a generalized Hausman Test, selecting between a GLS random effects (re) and a fixed effects model (fe). *,** and *** indicate statistical significance at the 10%, 5% and 1% levels, respectively.

with an $R^2$ of 13.1% for the WACC and 10% for the WACCCostofEquity. For the financial stakeholders, no significant effects exist concerning the WACCCostofDebt.

## 12.4 Rechecking with Theory

The category system of the analytical units is rechecked by analyzing its relevance in the data sample and identifying its link to existing theoretical findings in the sustainability reporting domain and concerning stakeholder theory.

Regarding the theoretical orientation of this research proposal, the theory does not "immerse the material"[51] in focus. On the contrary, it creates the structure and framework of the subject and unit of analysis. Involving research that focused on stakeholder orientation in the past allows for generating novel empirical insights with the help of existing theoretical contributions. Thereby, the content analysis enables to build the bridge between stakeholder orientation in a sustainability and corporate context with the help of a deductive research approach.

Initially developing from financial information addressed in annual reports to an audience of financial stakeholders such as creditors and shareholders, disclosed content in corporate reporting includes valuable non-financial information such as the orientation towards stakeholders being part of the social environment of a company.[52]

The first part of the empirical analysis shows that under the developments of a discursive change towards sustainability and the questioning of traditional business models, stakeholder orientation shows, first, robust results to be a significant and value-relevant factor for sustainability and financial performance, and second, a necessary consideration of what kind of stakeholders to be addressed in corporate reporting.

Lining up with recent studies that analyzed the influence of ESG disclosure on financial and sustainability performance on a broader scale,[53] the underlying empirical analysis focuses on a more fine-grained and focused aspect of corporate sustainability, i.e., stakeholder orientation.

Contrary to prior studies that could not find a direct link between social sustainability and its impact on the cost of capital[54], a more detailed view of the role of

[51] Mayring (2014), p. 59.

[52] see Owen (1990).

[53] see Dhaliwal et al. (2011), Eliwa et al. (2021), Y. Kim et al. (2012), Ng and Rezaee (2015), Pulino et al. (2022), Zumente and Bistrova (2021).

[54] see Ng and Rezaee (2015), p. 144.

stakeholders allows for further empirical interpretations. Therefore, the exploration joins the ranks of research that analyzed concrete stakeholders being part of corporate disclosure, such as Stocker et al. (2020), while further showing the relevance of stakeholder theory in the sustainability context in general and the analysis of social stakeholders as a relevant sustainable and financial value contributor in more detail.[55]

## 12.5 Summary and Interpretation of Results

Respecting the existing limitations of this dissertation, the empirical results show a robust signal that social stakeholder orientation is of utmost importance for companies. As the verified hypotheses H1a and H1b verify, when addressing social stakeholders in corporate reporting, namely the stakeholder society, a positive association with the sustainability rating Refinitiv ESG rating (RESGScore) exists.

Interpreting these results, growing corporate attention towards society increases the role of social stakeholders in defining the boundaries of organizational legitimacy, what is acceptable from a social reference point, and what companies have to justify. It further highlights how stakeholders that are linked to social issues become objective reality in business and for corporate sustainability reporting in more detail. Beyond the impact of social stakeholders on sustainability performance, its impact on financial performance, with hypothesis H2a, shows an additional degree of value relevance that also tackles the impact of stakeholders such as society, community, and employees on the firm's capital costs. Consequently, banks represent a stakeholder defining a company's cost of debt and considering the orientation toward social stakeholders as a relevant matter of financing risks. Parallel to these results, investors shaping a company's cost of equity also integrate companies' orientation towards social actors into their risk and performance considerations.

At the same time, as proven with the verification of hypothesis H2b, an inclination towards financial stakeholders in sustainability reporting reaches less appreciation by the financing costs of an entity. Due to a negative influence of the mentioning of the stakeholder bank on sustainability performance (RESGScore) and of the stakeholder shareholder on financial performance (WACC, WACCCostofEquity), the results reveal that the turf of sustainability reporting demands a more profound orientation towards social stakeholders, acknowledging and appreciating social stakeholder orientation by higher sustainability ratings and lower capital costs.

[55] see Stocker et al. (2020).

### 12.5.1 Implications for Regulators, Standard Setters, and Rating Agencies

The first part of the empirical investigation focusing on stakeholder orientation reveals the value relevance of social stakeholders such as society, community, and employees. It emphasizes the necessity to develop corporate disclosure standards that focus on the external environment of organizations and their most relevant players. Sustainability reporting opens the gate for a deeper consideration of social stakeholders to embrace the close link between social stakeholders and their impact on the people and the planet. Furthermore, in the sense of double materiality, its impact on financial aspects, such as the cost of capital, further underlines that social stakeholders matter for risk considerations of banks and investors.

Though being referenced for the first time in more detail as *social partners* and communities in the recently published CSRD of the EU, the understanding of the public as a relevant stakeholder still needs to be further incorporated in existing directives. Standard setters embracing the double materiality principle consider social stakeholders as a relevant and essential audience for corporate reporting. However, how to define and address social stakeholders and how to develop a sensitivity for its issues unveils further potential.

In addition, the results for the first hypothesis relating to social stakeholders can be interpreted in a way that rating agencies consider and appreciate the integration of stakeholders such as society, community, and employees as value-relevant stakeholders of corporate reporting. Besides its value relevance that is verified in this dissertation, how to include and address social stakeholders in corporate reporting, what rating agencies expect in terms of a social stakeholder orientation represent a crucial next step to learning more about the orientation of companies towards society and all other social stakeholders. Besides addressing social stakeholders in corporate reporting, the underlying issues that drive a corporate social-stakeholder relationship are the focus of the second part of the empirical investigation of this dissertation. Before diving into the succeeding analysis, the implications of the first analysis on companies, investors, and banks are part of the next section.

### 12.5.2 Implications for Companies, Investors, and Banks

The results of the first empirical analysis conclude that including social stakeholders in the corporate environment positively affects corporate sustainability performance. Companies are expected to include the social domain and its actors in their official communication and corporate disclosure. As a reference criterion for corporate risk and performance, social stakeholders not only reached the attention of existing standard setters and ESG rating agencies but also became a necessary addressee of companies that have to manage the expectations of their social environment. One central interpretation of the first part of empirical analysis is that social stakeholder orientation represents a crucial criterion factored into corporate evaluations by investors and banks. Banks link their credit facilitation to ESG risk attributes and sustainability ratings. Thus, lenders directly influence the borrower's ESG profile and performance, as explored by Houston and Shan (2022).[56] The same is true for investors and active owners who further engage with their portfolio companies on material ESG issues. If sustainability performance does not meet certain requirements, a direct effect on investments and their financial conditions is likely.[57].

The result of hypothesis H1b shows that the stakeholder bank, which is negatively associated with sustainability performance, does not expect an orientation towards its own but a decent examination of the corporate social environment. The same is true for the stakeholder shareholder. Speaking about the controversy between the concept of shareholder versus stakeholder value, developments towards sustainability in business and corporate reporting ask companies to include social stakeholders as a relevant value driver, moving from an inclined shareholder orientation to the risks and the opportunities that a social orientation of companies and distinguished social license of business operations bring. To reach the appreciation of rating agencies and to be able to guarantee solid financing of the business, companies have to consider the social environment and its players shaping the social discourse. Since corporate disclosure represents a result that makes corporate actions transparent, an increasing orientation toward social stakeholders asks companies to know about the expectations of social stakeholders and how to address them. In sum, the interpretation of the first results of this dissertation allows for the conclusion that, as social actors, companies have to align with their peers, establishing or preserving their organizational legitimacy to guarantee a going concern of the business or, at least, not putting a companies future at risk.

## 12.6 Application of Content Analysis II—Analyzing Public Value Topic Salience

The second part of the empirical research looks into the topic focus of corporate reporting with the help of the Sustainable Development Goals (SDGs). By analyzing corporate reporting with the help of a framework that considers environmental

[56] see Houston and Shan (2022).

[57] see Barko et al. (2022); Bauer et al. (2023); Dhaliwal et al. (2011).

and social topics, the topic salience that companies determine shall be further explored. Defining SDGs as content analytical units helps explore whether companies include the SDG framework and the single SDGs in their corporate disclosure. The descriptive statistics part explains how the companies in the data sample defined their focus on SDGs and how the focus may have shifted in the observed time frame between 2018 and 2021.

The inferential statistics part reveals how value-relevant the focus on different SDGs is and which public value categorizations on the public value landscape[58] play a dominant role in corporate disclosure. This helps to uncover relationships between sustainability topics that are the focus of global companies and the role these topics play in terms of sustainability and financial performance. It further enlightens a more detailed view of the relevance of public value as a value component for global businesses.

### 12.6.1 Deductive Categorization—Sustainable Development Goals as Content Analytical Units

Resting on the findings of the theory development in this dissertation, SDGs enable a focused view on social topics in corporate reporting while representing a framework that gains further attention and acceptance in research and business. To explore the societal discursive change that has happened in recent years that has a direct effect on the business world, the 17 SDGs are further consolidated into the ten goals that focus on society and social topics.

Though interactions between different SDGs exist and are explored by researchers,[59] this consolidation step is necessary to get a more detailed view of social topics and issues and their influence on sustainable and financial value relevance. Relating the clustering of SDGs to social topics as part of the deductive categorization, the analysis relates to the concept of the SDGs Wedding Cake of the Stockholm Resilience Center at Stockholm University.[60] Apart from SDG7 (Affordable Clean Energy) and SDG11 (Sustainable Cities and Communities) that are included in the SDGs Wedding Cake, the analysis concentrates on SDG 1, 2, 3, 4, 5 as SDGs related to society. SDG 8, 10, and 12 are added to the empirical investigation to analyze those with a social economic impact. In addition, SDG 16 and 17 are

[58] Meynhardt (2009).

[59] see Nilsson et al. (2018); Fonseca et al. (2020).

[60] Stockholm Resilience Center (2016).

further included with a focus on institutions and partnerships as social relationship criteria.

Consequently, the empirical investigation includes the following SDGs:

- SDG1 (No Poverty)
- SDG2 (Zero Hunger)
- SDG3 (Good Health and Well-Being)
- SDG4 (Quality Education)
- SDG5 (Gender Equality)
- SDG8 (Decent Work and Economic Growth)
- SDG10 (Reduced Inequalities)
- SDG12 (Responsible Consumption and Production)
- SDG16 (Peace, Justice and Strong Institutions)
- SDG17 (Partnerships for the Goals)

Thereby, the mentioned SDGs serve as content analytical units for exploring public value topic salience in the corporate reports of the data sample.

### 12.6.2 Quality Criteria

The second part of the empirical investigation is conducted manually since the SDG topic focus in corporate and sustainability reporting is often part of figure and image-related content. Automated text and information extraction related to the SDGs are at risk of not including SDG-related and relevant information published by the companies. Therefore, a manual approach is followed that searches for SDG mentions in the underlying text data. The analysis considers all sections that are part of corporate reports. The focus lies on the search for the SDGs in general to gather information on whether SDGs as a framework are in focus of a company and on single mentions of special SDGs to learn about topic salience and a specific SDG prioritization that a company follows.

The text number combination of linking the acronym SDG with an Arabic number between 1 and 17 of the content analytical units enables a manual exploration that is less influenced by linguistic and language-related specificities and, therefore, less prone to reliability problems such as stability, replicability, and accuracy of the analysis. As Krippendorff (2011) defined, "reliability is the extent to which [...] people arrive at the same interpretations of facts"[61].

[61] Krippendorff (2011), p. 94.; Bilolo (2018).

The public value landscape by Meynhardt (2009) builds the bridge between social topics in focus of corporate reporting and the legitimacy the companies strive for to guarantee their "social license to operate"[62]. Based on its clustering into moral-ethical, hedonistic-ethical, political-social, and utilitarian-instrumental categories, a suitable and well-accepted framework exists for analyzing public value topic salience in a corporate context.

### 12.6.3 Descriptive Statistics—Sustainable Development Goals in Corporate Reporting

Exploring SDGs in corporate reporting shall outline if social topics covering the issues of the SDG framework are salient subjects in corporate reporting. This section shows the average coverage based on the mention frequency of SDGs in the underlying sample and the cumulative change and growth of SDG mentions in the time frame between 2018 and 2021.

Based on the total mentions of SDGs, 1602 mentions, the average mention frequency over all SDGs is 40 mentions per year, with a single SDG reaching at least 21 mentions per year over the 358 reports that are published by the 90 companies included in the sample. Figure 12.7 shows the distribution of the single SDGs in the data sample.

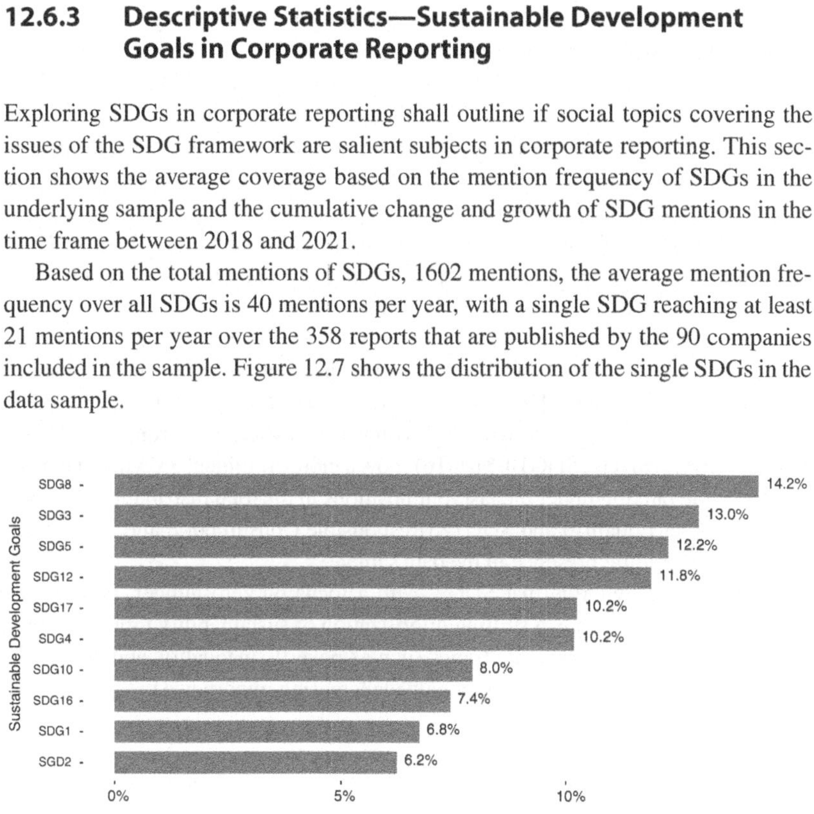


**Fig. 12.7** SDG Coverage in the Data Sample

[62] Marschlich and Ingenhoff (2021), p.1.

With a distribution of 14.2%, the most significant focus of the underlying corporate reports is on SDG8, which refers to decent work and economic growth. With 13%, the topic of good health and well-being (SDG3) is followed by SDG5, gender equality with 12.2%. Responsible consumption and production (SDG12) covers 11.8% of reports in the sample. The institutional SDG17 that focuses on partnerships to implement the goals, and the topic of quality education (SDG4) account for 10.2% of overall SDG mentions in the corporate reports. Reduced inequalities (SDG10) covers 8%, followed by peace, justice, and strong institutions (SDG16) and no poverty (SDG1) with 7.4% and 6.8%, respectively. An average share of 6.2% is attributed to SDG2 with the focus on zero hunger.

In sum, the distribution of SDGs in the data sample shows that social SDGs play a role in corporate reporting, being dominated with a cumulative share of over 50% by sustainable goals that relate to health and working conditions, to the role of gender equality and to the responsibility of producers and consumers that are addressed by SDG 8, 3, 5, and 12.

The cumulative changes in SDG mentions reveal a shift toward some SDGs that have been more neglected in corporate reporting as seen before. As Fig. 12.8 outlines, from 2018 until 2021, all SDGs reached an increase of mentions of at least 13% (SDG2). With a 52% growth in mentions, the highest growth rate is attributed to the topic of no poverty (SDG1) that, as the distribution shows in 12.7, ranked second last in the sample. The topic of reduced inequalities (SDG10) increased by 48%, followed by SDG16, which focuses on peace, justice, and strong institutions. All three SDGs (SDG1, SDG10, SDG16) show a more prominent growth trend in the sample while ranking under the last four positions in the total distribution. Besides, SDG5 (Gender Equality) and SDG12 (Responsible Consumption and Production) confirm their leading presence in the data sample.

Despite the minor role that SDG2, with a focus on zero hunger, plays in the reports of the data sample, a shifting prioritization of sustainability topics linked to the SDG framework is present. Topics such as poverty, inequality, and the role of peace, justice, and strong institutions gained increasing importance in the respective time frame.

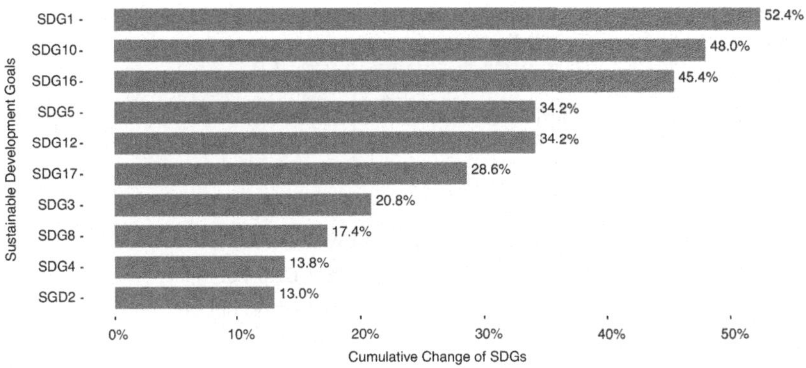


**Fig. 12.8** Cumulative Change of SDGs in the Data Sample

### 12.6.4 Inferential Statistics—The Value Relevance of Public Value Topic Salience

The inferential statistics of the second empirical analysis of this dissertation focuses on the regression of metrics that represent sustainable and financial performance with indicators that relate to the SDG framework, its issues, and topics. Before diving into the empirical exploration, the following section shows an overview of variables included in the analysis.

#### 12.6.4.1 Variables and Regression Models

The analysis of public value and its relationship with sustainability and financial performance relates to the variables mentioned in Table 12.9. The table shows the dependent variables, independent variables, and control variables that are part of the regression models for exploring public value topic salience.

**Table 12.9** List of Variables, Domains and Scores Applied in the Regression Model for Exploring Public Value Topic Salience

| Variables | Domain | Score |
|---|---|---|
| Dependant Variables | Sustainability Performance | BESScore |
| | | BSocScore |
| | | HumanRightsScore |
| | Financial Performance | WACC |
| | | WACCCost of Equity |
| | | WACCCost of Debt |
| Independent Variables | Public Value Topic Salience | SumSDG |
| | | SDG1, SDG2, SDG3, SDG4, |
| | | SDG5, SDG8, SDG10, SDG12 |
| | | SDG16, SDG17 |
| Control Variables | ESG-Related | RESGScore |
| | | SocialPillarScore |
| | | UNGCMembership |
| | | PolicyExecutiveCompensationScore |
| | Market-Related | lnRevenue |
| | Performance-Related | EBITDAMargin |

**Dependent Variables** The dependent variables in the underlying regression models include the Bloomberg Environmental and Social Score (BESScore), the Bloomberg Social Score (BSocScore), and the Refinitiv Human Rights Score (HumanRightsScore) for analyzing sustainability performance. For the exploration of financial performance, the Weighted Average Cost of Capital (WACC), including its two components, the cost of equity (WACCCostofEquity) and the cost of debt (WACCCostofDebt) are part of the regression models.

*BESScore*: Similar to the Refinitiv ESG metrics, the BESScore by Bloomberg measures sustainability performance. The ES Score components are the ESG headline score, ESG pillar score, theme scores, issue scores, sub-issue scores, and field scores. Integrated in the issue score is a disclosure factor that evaluates missing data not provided by the company. The materiality framework of Bloomberg includes global standards such as SASB and additional country-related guidelines. The total range of Bloomberg ES Scores goes from 0 to 10, while higher numbers indicate a better ESG performance rated by Bloomberg.[63] *BSocScore*: The variable BSocScore refers to the Bloomberg Social Pillar Score. Bloomberg's ESG Pillar Scores are a weighted mean of Issue Scores that are ranked according to the industries in focus. Bloomberg states that "the ranking of E and S pillars were assessed based on Bloomberg Intelligence analysis of the relative exposure to E and S issues identified in the industry's materiality framework"[64]. The Social Score goes from 0 to 10, while higher numbers indicate a better social performance rated by Bloomberg.

*HumanRightsScore*: The variable HumanRightsScore refers to the Refinitiv Human Rights Score and considers how companies take fundamental human rights conventions into account, expressed in absolute numbers ranging from 0 to 100. *WACC*: The variable Weighted Average Cost of Capital (WACC) is a metric for measuring a company's cost of capital as provided by Refinitiv. It included stocks (equity and preferred) and debt; each capital source is weighted proportionately and expressed as a percentage.

*WACCCostofEquity*: This variable refers to the weighted equity costs of the WACC. It refers to the risk premium that a company pays to its equity investors. The cost of equity results from the multiplied equity risk premium of the market combined with a company's beta factor of the stock. The inflation-adjusted risk-free rate is added, while the result is expressed as a percentage.

*WACCCostofDebt*: This variable refers to the weighted debt costs of the WACC. It includes the marginal cost for the company to issue new debt and the weighted cost of short-term and long-term debt, expressed as a percentage.

**Independent Variables** The independent variables in the regression model for analyzing stakeholder orientation contain the sum of SDGs (SumSDG) mentioned by companies being part of the data sample and the single social SDGs that are included in the regression models for analyzing public value topic salience.

*SumSDG*: The variable SumSDG represents the total number of SDGs mentioned per corporate report as an absolute number.

---

[63] see Bloomberg (n.d.).

[64] Bloomberg (n.d.), p. 15.

*SDG1*: This variable refers to the Sustainable Development Goal 1 in Boolean expression.

*SDG2*: This variable refers to the Sustainable Development Goal 2 in Boolean expression.

*SDG3*: This variable refers to the Sustainable Development Goal 3 in Boolean expression.

*SDG4*: This variable refers to the Sustainable Development Goal 4 in Boolean expression.

*SDG5*: This variable refers to the Sustainable Development Goal 5 in Boolean expression.

*SDG8*: This variable refers to the Sustainable Development Goal 8 in Boolean expression.

*SDG10*: This variable refers to the Sustainable Development Goal 10 in Boolean expression.

*SDG12*: This variable refers to the Sustainable Development Goal 12 in Boolean expression.

*SDG16*: This variable refers to the Sustainable Development Goal 16 in Boolean expression.

*SDG17*: This variable refers to the Sustainable Development Goal 17 in Boolean expression.

**Control Variables** The Refinitiv ESG Score and Social Pillar Score, the United Nations Global Compact Membership score (UNGCMembership) that refers to the status of a possible UNGC membership of the companies in focus, and the Policy Executive Compensation Score related to management compensation and its link to corporate ESG policies represent ESG-related control variables. Market-related variables such as revenues serve as controlling variables in the regression model. LnRevenue is considered with its natural logarithm to control for too much influence on the regression results of large revenues of the companies considered.

Further control variables included in the models are the performance-related metric EBITDAMargin.

*RESGScore*: The RESGScore refers to the Refinitiv ESG Score and entails around 630 company-level ESG metrics that are divided into ten different categories, illustrating "a company's ESG performance, commitment and effectiveness based on publicly-reported information"[65]. Every category, under consideration of its category weights, contributes to the ESG score. In comparison, the S and E pillar weights differ according to the industry in focus. The G pillar weighting does not vary. The pillar weights are summarized into an ESG score of 0–100. The D Score stands for poor relative ESG performance for ratings between 0 and 25. Excellent relative ESG performance, the A Score, is equal to scores ranging from 75 to 100.

*SocialPillarScore*: Being a part of the Refinitiv ESG Score, the SocialPillarScore relates to social subtopics that are covered in the Refinitiv ESG framework. Categories such as community, human rights, product responsibility, and workforce are addressed by the social pillars, covering additional themes listed under every category, such as product quality, data security, diversity and inclusion, and health and safety issues. Category weights are constructed with the help of an approach that considers its materiality per industry. As stated by Refinitiv, "data points with sufficient disclosure are used as a proxy for industry magnitude"[66]. The industry magnitude describes the relative weight defined by the relative industry median for a company and the transparency weight depending on the level of corporate disclosure in a specific industry group. As the ESG Score, the Social Pillar Score ranges from 0 to 100, while 100 equals the best sustainability score in that category rated by Refinitiv.[67]

*UNGCMembership*: This variable includes if a company is an official member of the United Nations Global Compact (UNGC) as stated on the official webpage of the United Nations Global Compact[68] in Boolean expression.[69]

---

[65] Refinitiv (2022), p. 6.

[66] Refinitiv (2022), p. 10.

[67] Refinitiv (2022), p. 7–11.

[68] United Nations (2023).

[69] This variable is only included in the analysis of financial performance and excluded from the analysis of sustainability performance to avoid multicollinearity since the status of UNGC membership is part of the rating methodology of certain third-party ESG rating providers.

*PolicyExecutiveCompensationScore*: This score represents an extra-financial performance component provided by Refinitiv and contains a management compensation factor related to ESG and sustainability in a range between 0 and 100.

*lnRevenue*: lnRevenue is the natural logarithm of the yearly revenues in a fiscal period of the specific company in focus, expressed in decimal form.

*EBITDAMargin*: The variable EBITDA Margin is the result of the yearly EBITDA in a fiscal period of the specific company in focus divided by its total revenue for the same period, expressed in percentage.

#### 12.6.4.2 Public Value Topic Salience and Sustainability Performance

With the help of the SDG framework and the included social SDGs, this section dives into the analysis of public value topic salience and its possible impact on sustainability performance. The exploration of SDGs as a public value component and a potential driver for sustainability performance is based on the Bloomberg ESG Score (BESScore) and the Social Score (BSocScore), and the score for analyzing human rights alignment (HumanRightsScore), which serve as dependent variables. The analysis includes the single SDGs and the sum of SDGs (SumSDG) as independent variables. It further considers the mentioned control variables to control for further influence on the models.

The following hypothesis states an expected result for exploring public value topic salience and its impact on sustainability performance.

*H3 Public value topic salience is positively associated with sustainability performance.*

The following regression model 14 serves to verify and assess the stated hypothesis H3 based on the mentioned variables.[70]

$$Model(14)$$

$$BESScore_{i,t} = \alpha_i + \beta_1[SumSDG]_{i,t} + \beta_2[RESGScore]_{i,t} + \beta_3[UNGCMembership] + \beta_4[PolicyExecutiveCompensationScore]_{i,t} + \beta_5[EBITDAMargin]_{i,t} + u_{i,t}$$

[70] The market-related control variables were excluded due to a high multi-collinearity with the variables of the model while still showing high significance.

Under consideration of the control variables RefinitivESGScore, UNGCMembership, PolicyEcecutiveCompensationScore, and the EBITDAMargins of the single companies that validate the model, Model 14 aims to explore the relationship of the sum of SDGs disclosed and the Bloomberg Environmental and Social Score. As Table 12.10 reveals, with a very high significance level below 1% and an $R^2$ of 23.7%, hypothesis H3 can be verified, showing a robust significance and positive correlation between the sum of SDGs reported and the sustainability performance rating (BESScore). The control variables RESGScore, UNGCMembership, and the EBITDAMargin also show a high significant level below 1%, while the PolicyExecutiveCompensationScore is below 10%.

**Table 12.10** The Correlation of the sum of SDGs With the Bloomberg ES Rating

| MODEL (fe) | (14) |
|---|---|
| VARIABLES | BESScore |
| SumSDG | 0.0399*** |
| | (0.00966) |
| RESGScore | 0.0358*** |
| | (0.0114) |
| UNGCMembership | 0.842*** |
| | (0.0637) |
| PolicyExecutiveCompensationScore | −0.214* |
| | (0.116) |
| EBITDAMargin | 2.110*** |
| | (0.756) |
| Constant | 0.489 |
| | (0.876) |
| Observations | 349 |
| R-squared (within) | 0.237 |
| Number of companycode | 88 |

*Note.* All models are estimated as robust standard errors (in parentheses). Model selection is based on a generalized Hausman Test, selecting between a GLS random effects (re) and a fixed effects model (fe). *,** and *** indicate statistical significance at the 10%, 5% and 1% levels, respectively.

Showing that the sum of SDGs has a positive and significant effect on sustainability performance ratings, it remains the question of which SDGs show the highest correlation factors. Therefore, Model 15 is considered to dive deeper into the analysis of the single social SDGs.

$$Model(15)$$

$$BESScore_{i,t} = \alpha_i + \beta_1[SingleSDG]_{i,t} + \beta_2[RESGScore]_{i,t} + \beta_3[UNGCMembership]_{i,t} + \beta_4[PolicyExecutiveCompensationScore]_{i,t} + \beta_5[EBITDAMargin]_{i,t} + u_{i,t}$$

As Table 12.11 and 12.12 reveal, all social SDGs, except SDG2 (Zero Hunger), are statistically significant. Apart from SDG4 with a significance level below 10% and SDG1 below 5%, all other significance levels are below the 1% level, showing the highest correlation factors for SDG10 (Reduced Inequalities), followed by SDG5 (Gender Equality), SDG12 (Responsible Consumption and Production), SDG3 (Good Health and Well-Being), SDG1 (No Poverty), SDG17 (Partnership for the Goals), SDG16 (Peace and Justice), SDG8 (Good Jobs and Economic Growth), and SDG4 (Quality Education). Consequently, social SDGs indicate a significant effect on the high-level sustainability performance rating and therefore verify hypothesis H3.

**Table 12.11** The Correlation of Single SDGs With the Bloomberg ES Rating (BESScore)

| Variables | BESScore | BESScore | BESScore | BESScore | BESScore |
|---|---|---|---|---|---|
| RESGScore | 0.0380*** | 0.0367*** | 0.0399*** | 0.0365*** | 0.0376*** |
| | (0.0114) | (0.0117) | (0.0122) | (0.0117) | (0.0115) |
| UNGCMembership | 0.931*** | 0.922*** | 0.984*** | 0.910*** | 0.946*** |
| | (0.0630) | (0.0572) | (0.0660) | (0.0613) | (0.0540) |
| PolicyExecutive | −0.247* | −0.235* | −0.264** | −0.211* | −0.231* |
| CompensationScore | (0.124) | (0.119) | (0.121) | (0.120) | (0.119) |
| EBITDAMargin | 2.090** | 2.025*** | 2.043** | 2.059*** | 2.115*** |
| | (0.825) | (0.754) | (0.791) | (0.746) | (0.769) |
| SDG | [1] 0.236** | [3] 0.243*** | [4] 0.143* | [5] 0.267*** | [8] 0.204*** |
| | (0.0990) | (0.0727) | (0.0815) | (0.0843) | (0.0754) |
| Constant | 0.410 | 0.448 | 0.264 | 0.441 | 0.353 |
| | (0.871) | (0.898) | (0.933) | (0.897) | (0.883) |
| Observations | 349 | 349 | 349 | 349 | 349 |
| R-squared (within) | 0.217 | 0.223 | 0.206 | 0.231 | 0.215 |
| Number of companycode | 88 | 88 | 88 | 88 | 88 |

*Note.* All models are estimated as robust standard errors (in parentheses). The SDG in focus is highlighted in square brackets. Model selection is based on a generalized Hausman Test, selecting between a GLS random effects (re) and a fixed effects model (fe). *,** and *** indicate statistical significance at the 10%, 5% and 1% levels, respectively.

**Table 12.12** The Correlation of Single SDGs With the Bloomberg ES Rating (BESScore)

| Variables | BESScore | BESScore | BESScore | BESScore |
|---|---|---|---|---|
| RESGScore | 0.0346*** | 0.0363*** | 0.0376*** | 0.0381*** |
| | (0.00999) | (0.0117) | (0.0118) | (0.0120) |
| UNGCMembership | 0.828*** | 0.919*** | 0.944*** | 0.938*** |
| | (0.0543) | (0.0538) | (0.0566) | (0.0603) |
| PolicyExecutive | −0.200** | −0.216* | −0.250** | −0.245** |
| CompensationScore | (0.0924) | (0.119) | (0.117) | (0.117) |
| EBITDAMargin | 2.019** | 2.192*** | 2.066** | 2.058*** |
| | (0.806) | (0.806) | (0.808) | (0.777) |
| SDG | [10] 0.416*** | [12] 0.251*** | [16] 0.206*** | [17] 0.221*** |
| | (0.0825) | (0.0672) | (0.0663) | (0.0783) |
| Constant | 0.631 | 0.425 | 0.441 | 0.373 |
| | (0.759) | (0.899) | (0.902) | (0.918) |
| Observations | 349 | 349 | 349 | 349 |
| R-squared (within) | 0.269 | 0.228 | 0.214 | 0.222 |
| Number of companycode | 88 | 88 | 88 | 88 |

*Note.* All models are estimated as robust standard errors (in parentheses). The SDG in focus is highlighted in square brackets. Model selection is based on a generalized Hausman Test, selecting between a GLS random effects (re) and a fixed effects model (fe). *,** and *** indicate statistical significance at the 10%, 5% and 1% levels, respectively.

Since the BESScore, besides social topics, also includes environmental issues rated by the rating providers, the impact on social performance is the direct focus of Model 16.

$$Model(16)$$

$$BSocScore_{i,t} = \alpha_i + \beta_1[SumSDG]_{i,t} + \beta_2[SocialPillarScore]_{i,t} + \beta_3[EBITDAMargin]_{i,t} + u_{i,t}$$

As Table 12.13 shows, with a significance below the 1% level and with an effect size of 12.7% ($R^2$), H3 can also be verified for the social performance rating (BSocScore) showing a positive correlation between the sum of SDGs reported and the Bloomberg

**Table 12.13** The Correlation of the sum of SDGs With the Social Performance Score BSocScore

| MODEL (fe) | (16) |
|---|---|
| VARIABLES | BSocScore |
| SumSDG | 0.0493*** |
| | (0.0148) |
| SocialPillarScore | 0.0351** |
| | (0.0142) |
| EBITDAMargin | 2.937** |
| | (1.292) |
| Constant | −0.692 |
| | (1.158) |
| Observations | 358 |
| R-squared (within) | 0.127 |
| Number of companycode | 90 |

*Note.* All models are estimated as robust standard errors (in parentheses). Model selection is based on a generalized Hausman Test, selecting between a GLS random effects (re) and a fixed effects model (fe). *,** and *** indicate statistical significance at the 10%, 5% and 1% levels, respectively.

Social Score by taking the significant control variables of the Refinitiv Social Pillar Score (SocialPillarScore) and the EBITDAMargin into account.[71]

However, which single SDGs indicate an influence on the social performance rating? Though having to be cautious due to lower scores for the goodness of fit of ($R^2$),

$$Model(17)$$

$$BSocScore_{i,t} = \alpha_i + \beta_1[SingleSDG]_{i,t} + \beta_2[SocialPillarScore]_{i,t} +$$

$$\beta_3[EBITDAMargin]_{i,t} + u_{i,t}$$

Model 17 in Table 12.14 and Table 12.15 shows that apart from SDG2 (Zero Hunger) and SDG4 (Quality Education), all other social SDGs show significant results. According to the correlation factors, SDG10 (Reduced Inequalities) shows the highest correlation, followed by SDG3 (Good Health and Well-Being), SDG16 (Peace, Justice and Strong Institutions), SDG17 (Partnerships for the Goals), SDG8 (Decent Work and Economic Growth), SDG1 (No Poverty), SDG12 (Responsible Consumption and Production), and SDG5 (Gender Equality).

[71] Additional control variables were excluded due to non-significance.

**Table 12.14** The Correlation of Single SDGs With the Social Performance Score BSocScore (part I)

| Variables | BSocScore | BSocScore | BSocScore | BSocScore |
|---|---|---|---|---|
| SocialPillar | 0.0298*** | 0.0278*** | 0.0296*** | 0.0295*** |
| Score | (0.00749) | (0.00741) | (0.00763) | (0.00743) |
| EBITDA | 1.741* | 1.658** | 1.713** | 1.740** |
| Margin | (0.894) | (0.829) | (0.860) | (0.851) |
| SDG | [1] 0.301** | [3] 0.385*** | [5] 0.253** | [8] 0.326*** |
| | (0.146) | (0.100) | (0.105) | (0.108) |
| Constant | 0.203 | 0.254 | 0.178 | 0.113 |
| | (0.639) | (0.626) | (0.638) | (0.632) |
| Observations | 358 | 358 | 358 | 358 |
| R-squared (between) | 0.0612 | 0.0903 | 0.061 | 0.0779 |
| Number of companycode | 90 | 90 | 90 | 90 |

*Note.* All models are estimated as robust standard errors (in parentheses). The SDG in focus is highlighted in square brackets. Model selection is based on a generalized Hausman Test, selecting between a GLS random effects (re) and a fixed effects model (fe). *,** and *** indicate statistical significance at the 10%, 5% and 1% levels, respectively.

**Table 12.15** The Correlation of Single SDGs With the Social Performance Score BSocScore (part II)

| Variables | BSocScore | BSocScore | BSocScore | BSocScore |
|---|---|---|---|---|
| SocialPillar | 0.0294*** | 0.0301*** | 0.0292*** | 0.0287*** |
| Score | (0.00730) | (0.00807) | (0.00711) | (0.00742) |
| EBITDA | 1.680* | 1.794** | 1.696* | 1.665** |
| Margin | (0.890) | (0.887) | (0.878) | (0.846) |
| SDG | [10] 0.500*** | [12] 0.265** | [16] 0.349*** | [17] 0.340*** |
| | (0.135) | (0.103) | (0.109) | (0.117) |
| Constant | 0.168 | 0.114 | 0.238 | 0.245 |
| | (0.624) | (0.659) | (0.603) | (0.615) |
| Observations | 358 | 358 | 358 | 358 |
| R-squared (between) | 0.0691 | 0.0645 | 0.0634 | 0.0743 |
| Number of companycode | 90 | 90 | 90 | 90 |

*Note.* All models are estimated as robust standard errors (in parentheses). The SDG in focus is highlighted in square brackets. Model selection is based on a generalized Hausman Test, selecting between a GLS random effects (re) and a fixed effects model (fe). *,** and *** indicate statistical significance at the 10%, 5% and 1% levels, respectively.

Consequently, similar to the influence on the general ESG rating (BESScore), the effect on the social performance rating (BSocScore) is dominated by SDG10 tackling reduced inequalities. SDG3 for good health and well-being correlates similarly with the social performance rating. At the same time, the cooperation SDGs 16 and 17 have higher effects on the social performance rating than the general sustainability score. SDG5 (Gender Equality), which shows high correlation factors for the overall rating, ranks last though still being significantly below a 5% significance level in the regression model focusing on social performance.

From social performance rating to an even more detailed view of an essential subtopic in the social domain, human rights alignment is further considered and tested in Model 18. However, the sum of SDGs shows no significant correlation below the 10% level on the HumanRightsScore.

$$Model(18)$$

$$HumanRightsScore_{i,t} = \alpha_i + \beta_1[SumSDG]_{i,t} + \beta_2[BSocScore]_{i,t} + \beta_3[UNGCMembership]_{i,t} + u_{i,t}$$

Therefore, taking a closer look at the single SDGs as stated in Model 19 and Table 12.16, SDG1 (No Poverty) and SDG4 (Quality Education) indicate an influence and positive correlation with the Refinitiv HumanRightsScore.

$$Model(19)$$

$$HumanRightsScore_{i,t} = \alpha_i + \beta_1[SingleSDG]_{i,t} + \beta_2[BSocScore]_{i,t} + \beta_3[UNGCMembership]_{i,t} + u_{i,t}$$

SDG4 shows a higher significance level of below 5% and a higher correlation factor, while SDG1 is below a 10% significant level. Both variables show a solid goodness of fit of 25.28% (SDG1) and 25.97% (SDG4). Consequently, different from the overall ESG rating (BESScore) and the social performance rating (BSocScore), where SDG4 showed lower significance levels, the impact of the SDG4 related to quality education seems to matter for the appraisal of how companies take the UN Human Rights Convention into account. As expected, SDG1 covering measures against poverty is even more dominant in the human rights rating compared to the overall sustainability and the social sustainability performance rating.

**Table 12.16** The Correlation of the Single SDGs With the HumanRightsScore

| Variables | HumanRightsScore | HumanRightsScore |
|---|---|---|
| BSocScore | 2.162*** | 2.184*** |
| | (0.741) | (0.743) |
| UNGCMembership | 21.47*** | 21.25*** |
| | (3.906) | (3.887) |
| SDG1 | 2.742* | |
| | (1.408) | |
| SDG4 | | 3.227** |
| | | (1.430) |
| Constant | 52.01*** | 51.43*** |
| | (4.999) | (5.031) |
| Observations | 358 | 358 |
| R-squared (between) | 0.2528 | 0.2597 |
| Number of companycode | 90 | 90 |

*Note.* All models are estimated as robust standard errors (in parentheses). Model selection is based on a generalized Hausman Test, selecting between a GLS random effects (re) and a fixed effects model (fe). *,** and *** indicate statistical significance at the 10%, 5% and 1% levels, respectively.

#### 12.6.4.3 Public Value Topic Salience and Financial Performance

To explore public value topic salience and to verify its effect on financial performance, the following hypothesis is defined.

*H4 Public value topic salience is negatively associated with the cost of capital.*

For the variable SumSDG, no significant correlation can be detected with the cost of capital. However, the following model helps to explore the relationship and a possible significant correlation between the single SDGs and the cost of capital. It includes the variable of the membership status of the single companies in the sample of the UN Global Compact (UNGCMembership), the variable PolicyExecutiveCompensationScore that controls for the influence of ESG-related performance compensation, and the performance- and market-related control variables EBITDAMargin and lnRevenue.

All single social SDGs are tested, while only SDG4 shows a significant relationship and is therefore embedded in model 17 and model 18 for showing its significant relationship with the WACC and the WACCCostofEquity.

$$Model(20)$$

$$WACC_{i,t} = \alpha_i + \beta_1[SDG4]_{i,t} + \beta_2[UNGCMembership]_{i,t} + \beta_3[PolicyExecutiveCompensationScore]_{i,t} + \beta_4[EBITDAMargin]_{i,t} + \beta_5[lnRevenue]_{i,t} + u_{i,t}$$

$$Model(21)$$

$$WACCCostofEquity_{i,t} = \alpha_i + \beta_1[SDG4]_{i,t} + \beta_2[UNGCMembership]_{i,t} + \beta_3[PolicyExecutiveCompensationScore]_{i,t} + \beta_4[EBITDAMargin]_{i,t} + \beta_5[lnRevenue]_{i,t} + u_{i,t}$$

Table 12.17 verifies the negative association between SDG4 and the WACC, resulting in a positive effect of a topic focus on the SDG that considers Quality Education.

**Table 12.17** The Correlation of SDG4 With the Cost of Capital (WACC, WACCCostofEquity)

| MODEL (fe) | (20) | (21) |
|---|---|---|
| VARIABLES | WACC | WACCCostofEquity |
| SDG4 | −0.00606* | −0.00840* |
| | (0.00352) | (0.00448) |
| UNGCMembership | −0.0256*** | −0.0231*** |
| | (0.00179) | (0.00234) |
| PolicyExecutiveCompensationScore | −0.0117** | −0.0122** |
| | (0.00482) | (0.00481) |
| EBITDAMargin | 0.0830** | 0.0900* |
| | (0.0337) | (0.0475) |
| lnRevenue | 0.0287*** | 0.0268*** |
| | (0.00805) | (0.0100) |
| Constant | −0.625*** | −0.571** |
| | (0.191) | (0.237) |
| Observations | 358 | 358 |
| R-squared (within) | 0.132 | 0.101 |
| Number of companycode | 90 | 90 |

*Note.* All models are estimated as robust standard errors (in parentheses). Model selection is based on a generalized Hausman Test, selecting between a GLS random effects (re) and a fixed effects model (fe). *,** and *** indicate statistical significance at the 10%, 5% and 1% levels, respectively.

#### 12.6.4.4 Public Value Topic Salience on the Public Value Landscape

To understand how to classify and categorize the results of the empirical investigation according to their relationship with public value, the public value landscape by Meynhardt (2009) provides a suitable structuring framework. First, it shapes the understanding of the single SDGs regarding which type of public value dimension is targeted. Second, the categorization allows us to explore the impact of the four public value dimensions on sustainable and financial performance to draw further conclusions. Therefore, Fig. 12.9 shows the single social SDGs clustered among the public value landscape.

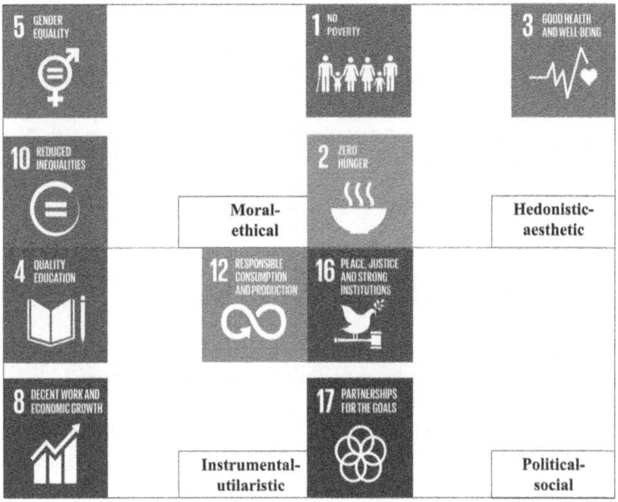


**Fig. 12.9** SDG Classification on the Public Value Landscape by Meynhardt 2009 (Personal Communication, February 13, 2023)

SDG 5 and 10 relate to the moral-ethical category evaluating organizational actions to reduce general inequalities and supporting gender equality, referring to what is "seen just or unjust in a certain social context"[72].

Out of the perspective of organizations, SDG 1, 2, and 3 focus on the hedonistic-aesthetic category due to the actions in focus that are contributing to individual health and well-being, considering individual needs and the fight against poverty and hunger.

[72] Meynhardt (2015), p. 153.

SDG 4, 8, and 12, with their focus on quality education, decent work and economic growth, and responsible consumption and production, relate to the instrumental-utilaristic category since this category deals with the understanding of the world, its conceptions of an economy and how organizations position and orientate themselves in the current environment.[73]

SDG 16 and 17 are part of the political-social category involving external institutions, partnerships, relationships, and its qualities.

As a next step, the single SDGs are clustered on the public value landscape and added with its correlation factors towards the sustainability performance score BESSccore.

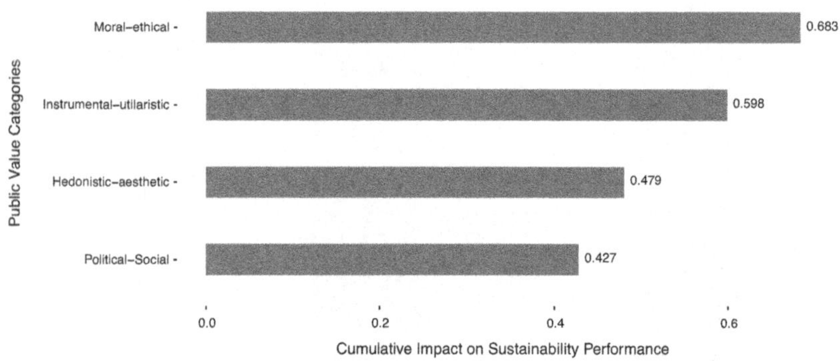


**Fig. 12.10** Public Value Categories and Their Impact on Sustainability Performance

Figure 12.10 reveals that the moral-ethical dimension (0.683) with SDG 5 and 10 that refer to gender equality and reduced inequalities have the highest impact on sustainability performance, followed by the instrumental-utilaristic dimension (0.598) with SDG 4, 8, and 12 that focus on quality education, decent work and economic growth, and on responsible consumption and production. The hedonistic-aesthetic dimension (0.479) and the political-social dimension (0.427) do also show significant results, ranking third and fourth in terms of their correlation with sustainability performance in comparison with the two other public value dimensions.

As a next step, all SDGs are clustered on the public value landscape and added with its correlation factors towards the social sustainability performance score BSoc-Score analyzing the relationship with social performance.

[73] Meynhardt (2015), p. 154.

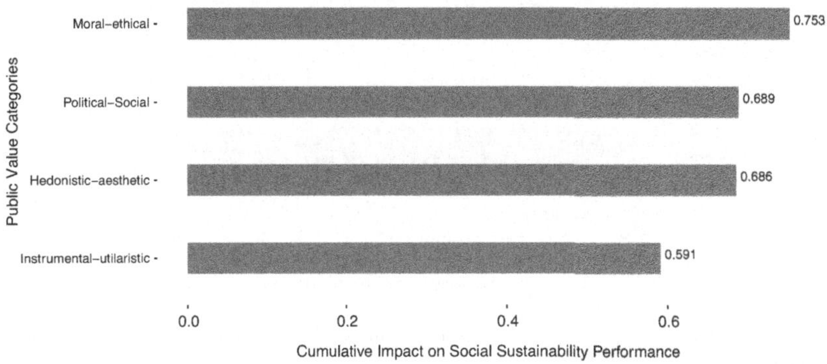


**Fig. 12.11** Public Value Categories and Their Impact on Social Sustainability Performance

Figure 12.11 shows that parallel to the impact on general sustainability performance that includes the view on environmental and social topics with the BESScore, the focused perspective on the social pillar emphasizes the role of the moral-ethical dimension (0.753) with SDG 5 and 10. However, the political-social dimension (0.689) ranking second reveals a different picture since SDG 16 and 17, with its focus on institutions, partnerships, and relationships, are of higher importance for the social evaluation of companies in sustainability ratings compared to the previous analysis. Compared to the greater importance of the instrumental-utilaristic dimension (0.591) that showed more significant correlation factors with the BESScore, its relationship with the social BSocScore shows lower correlation factors, ranking fourth behind the hedonistic-aesthetic dimension (0.686).

For the alignment with the Human Rights Convention and the HumanRightsScore, the following Fig. 12.12 only shows significant results for the instrumental-utilaristic dimension (3.227) including SDG 4, 8, and 12 with its focus on quality education, decent economic growth and the responsibility of producers and consumers and the hedonistic-aesthetic dimension (0.2742) that covers SDG 1, 2, and 3 with more obvious human rights related issues such as no poverty (SDG1), zero hunger (SDG2), and good health and well-being (SDG3).

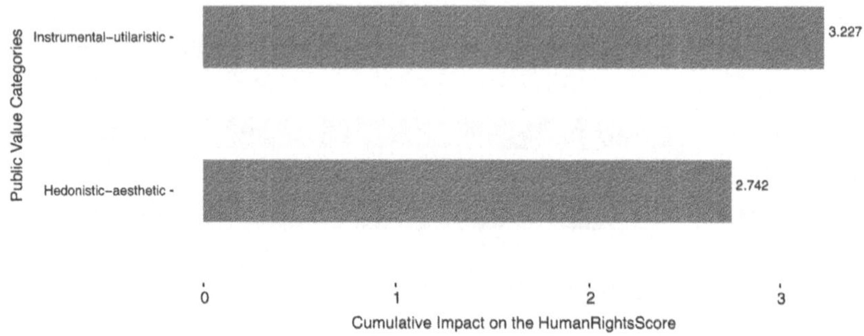


**Fig. 12.12** Public Value Categories and Their Impact on the Appraisal of Human Rights Alignment

For exploring the impact of the public value categories on financial performance, only SDG 4 (Quality Education) relating to the instrumental-utilaristic category, as shown in Table 12.17, reveals its influence on the WACC and the WACCCostofEquity. No significant correlation can be attributed to the WACCCostofDebt. Nevertheless, the instrumental-utilaristic category implies the understanding of the world focusing on knowledge and education and is thus a factor that is closely related to financial performance.

## 12.7 Rechecking with Theory

Criticisms concerning the SDGs relate to its ideology and governance concerns.[74] Furthermore, researchers acknowledged a lack of accountability and transparency with its link to accounting practices.[75] In addition, Bierman et al. (2022), in their assessment of over 3000 studies, concluded that transformational aspects of the SDGs for sustainable development after its proclamation in 2015 until today are still limited.[76]

However, the number of studies that recently focused on the scientific exploration of SDGs and the greater inclusion of the SDGs into sustainability reporting, as Curtó-Pagès et al. (2021) revealed, are a sign of its growing importance in the

[74] For governance concerns concerning the SDGs, see B. Kim et al. (2016); for its ideological concerns, see Nilsen (2016); H. Weber (2017).

[75] see Bebbington and Unerman (2018); Biermann et al. (2017).

[76] Biermann et al. (2022).

corporate context.[77] Reflecting an organization's business model and strategy with the understanding of sustainable development as understood by the SDGs became an additional task and duty of organizations.

This is also shown by the fact that the single SDGs in the data sample and in the underlying time frame grew at least 13%, speaking in favor of studies that speak of a greater alignment of indicators reported that are covered by the SDGs as shown in studies such as Perello-Marin et al. (2022) and Szennay et al. (2019).[78]

A verification of the impact of societal discourse and its shifts with its relevant influence on sustainability and financial performance that lets the understanding of the SDGs become a point of reference for organizational legitimacy is subject to the interpretation of the results in the following section.

## 12.8 Summary and Interpretation of Results

The results of the empirical analysis exploring the impact of public value topic salience on sustainability and financial performance show that the SDG framework is of relevance in the context of corporate sustainability disclosure. Created initially to focus on a country-based level, SDGs are nowadays the focus of companies challenged by the question of how to integrate the framework into corporate disclosure and related actions. Furthermore, the analysis goes beyond existing studies that, though contributing to the understanding of SDGs in corporate reporting, only concentrate on descriptive statistics on which and how SDGs are part of corporate reporting without allowing any interpretation and conclusion on its value relevance.

The results of the regression models reveal that SDG disclosure as a total of SDGs reported (SumSDG) indicates an impact on sustainability performance and, on a more detailed level, on social performance. The impact on financial performance and the rating that considers the Human Rights Convention shows no statistical significance for the variable that considers the sum of SDGs. With a corporate lens, this asks for a more granular view of the single SDGs in focus.

Apart from SDG2, the analysis of the single SDGs shows that there is a high and robust significance of almost all social SDGs on the sustainability performance rating (BESScore). With a further exclusion due to the non-significance of SDG4, the remaining eight SDGs also show a relevant correlation with a more detailed focus on social sustainability performance (BSocScore). However, the influence of the single SDGs on the general ESG Score (BESScore) differs in terms of the correlation fac-

[77] see Curtó-Pagès et al. (2021).

[78] Perello-Marin et al. (2022); Szennay et al. (2019).

tor compared to the impact on social sustainability performance (BSocScore). With the most remarkable correlation of SDG10 tackling inequalities for both the general sustainability and social performance rating, SDG5 targeting gender equality, followed by responsible consumption and production (SDG12), and good health and well-being (SDG3) show the most significant correlation on the BESScore. While SDG3 is also highly correlated with the social performance rating (BSocScore), the goals SDG 16 and 17 ask for targeting peace, justice, and the support of strong institutions and partnerships to realize the SDGs are ranking higher compared to the general sustainability rating.

This asks for a differentiated view of companies on how serious sustainability goals can be targeted and what priorities in the corporate agenda are set to follow prioritized goals and related actions. Though such a quantitative exploration can only show the significance of sustainability targets of companies on a very detailed and focused scale by considering subjective corporate disclosure, the discursive change that takes place concerning the SDGs and the relevance of public value on sustainability and financial performance is proven.

Consequently, compared to all other social SDGs, SDG4 plays a minor role in the social performance rating (BSocScore). It has no role in the general sustainability rating in focus (RESGScore) but, together with SDG1, is dominant in the regression with the HumanRightsScore. Quality education in the SDG framework, listed as goal number 4, is a central human right mentioned in Article 26 of the Universal Declaration of Human Rights, which states that

> "Everyone has the right to education. Education shall be free, at least in the elementary and fundamental stages. Elementary education shall be compulsory. Technical and professional education shall be made generally available and higher education shall be equally accessible to all on the basis of merit"[79].

Intriguingly, SDG4 does not only show the highest correlation factors for the pursuance of the Human Rights Convention but also in its effect on the financial performance ratings, the WACC, and the WACCCostofEquity. It reveals that members of society have significant responsibilities to enable a qualitative and sophisticated approach to education after guaranteeing general access and building an education infrastructure. The analysis stipulates that corporations seize opportunities and acknowledge their responsibility to foster education. As an interpretation of the results, education does not only have a favorite effect on the well-being of society by fostering the development of human rights for all individuals but, at the same time,

---

[79] United Nations (n.d.), p. 7.

reaches appreciation by investors in the form of the lower general cost of capital and equity financing as part of its financial performance.

The analysis of the public value dimensions on the public value landscape shows that the moral-ethical dimension, including SDG 5 and 10, reveals the highest impact on sustainability and social performance. It aligns with existing studies such as Hildebrand (2021) that highlighted that the cost of inaction as a share of global GDP is the highest for SDG5, compared to all other 17 SDGs.[80] However, SDGs related to the moral-ethical dimension are not understood as a significant materiality indicator, at least not in the sense of financial materiality provided by SASB, which asks standard setters to dedicate further efforts to standards that tackle this value dimension.

### 12.8.1 Implications for Regulators, Standard Setters, and Rating Agencies

Similar to the first empirical investigation and its implications for standard setters, the empirical results of the second part focusing on social issues and public value topic salience prove that a deeper consideration of existing frameworks such as the SDGs, the UNGC, and the concept of basic needs and public value are worth considering for the advancing drafts and its adaptations in the field of corporate sustainability disclosure.

Embracing societal discourse as a reference for disclosure requirements in the social domain might advance corporate disclosure, anticipate issues, and help standard setters to set the proper focus in the social pillar of sustainability.

For example, it is of high interest that SDG4, focusing on quality education, among the ten social SDGs in focus, is the only SDG showing significance on financial performance in the underlying data sample. Contrary to the results of this dissertation, Hildebrand (2021) analyzed that with 0%, SDG4 ranks last in its alignment with materiality indicators provided by SASB, which asks for a reassessment of SDG4 and how to include it in financial materiality assessments of standard setters.[81]

The same is true for reflections about how double materiality can be further defined and fostered in sustainability standards. The empirical analysis reveals that the sum of SDGs and most of the single SDGs show a robust, significant, and positive association with sustainability performance. It should catch the attention of stan-

[80] Hildebrand et al. (2021).

[81] Hildebrand et al. (2021).

dards setters to discuss how the SDG framework, its subtopics, and indicators can be further included in sustainability reporting standards. As initiated by Adams (2020, 2021), standard setters should also address how companies can better disclose SDG information and how double materiality analysis can be tackled more structuredly and efficiently.[82]

For rating agencies, further efforts are necessary to tackle the understanding of the moral-ethical dimension, its indicators, and performance targets. Since SDG 5 and 10, which refer to gender equality and reduced inequalities, have the highest impact on sustainability and social performance, further dialogue is required to create broader knowledge about these goals and their related issues and how companies can better address such basic needs of society. Besides, the social performance dimension that isolates the S pillar from additional environmental and governance considerations clearly shows significance for SDG 16 and 17, referring to the political-social dimension, asking for peace, justice, and stronger institutions (SDG16) as well as for partnerships for reaching the goals (SDG17). Therefore, standard setters and rating agencies could showcase best practices in this regard.

With a focus on human rights, further aspirations to define the impact of the instrumental-utilaristic value dimension covering SDG 4, 8, and 12 with its lens on quality education, decent economic growth, and the responsibility of producers and consumers ask for clearer guidelines, as recently pushed by regulations in the EU and Switzerland. The hedonistic-aesthetic dimension that covers SDG 1, 2, and 3 with more obvious human rights-related issues such as no poverty (SDG1), zero hunger(SDG2), and good health and well-being (SDG3) also requires targets, on a sectoral level, on how to better implement these goals into corporate reporting and from a rating perspective, how performance can be more accurately defined concerning these targets.

### 12.8.2 Implications for Companies, Investors, and Banks

Companies and investors are challenged to dive deeper into the exploration of social issues and their performance indicators related to business models and general corporate activities. As the results of the empirical investigation show, embracing the SDG framework, its subgoals, and its link with public value dimensions can support companies in their corporate disclosures and actions focusing on social issues. Companies can be motivated and intrigued to see that it is worth addressing social issues, first, for the sake of humanity to take responsibility as a social actor, and second,

[82] see Adams et al. (2020); Adams et al. (2021).

for reaching a higher sustainability performance in terms of a better sustainability rating, and better financial performance in terms of lower capital costs.

The results further highlight that the need for tackling social issues as a company is not only a result of stricter sustainability regulations, such as the EU CSRD, but also a necessity to comply with discursive shifts that happen in society that expect companies to prove organizational legitimacy employing a transparent approach towards sustainability reporting in general, and by addressing social issues in more detail. If a more significant corporate positioning on social issues is part of societal discourse with the potential to become a *Weltanschauung* and if the related knowledge about these issues and related actions are expected from companies, companies will feel the pressure regarding their organizational legitimacy if a clear and transparent communication on its social contributions is lacking. Questions will be addressed by society and communities, and further tackle the interest of other stakeholders such as employees, customers, banks, and investors.

At the same time, as the analysis confirms, investors are becoming more active in influencing corporate sustainability agendas. The significant results related to the positive association of SDG4 with the cost of equity reveal that a strategy on how to tackle salient public value topics such as fostering quality education is appreciated by investors, which might lead to greater sophistication and competition on the investor side, uncovering the financial and sustainable value of the social domain. As literature shows, more investors focus on ESG engagements if sustainability performance metrics do not satisfy the ownership.[83] The same holds true for banks. A greater understanding of their own sustainability controversies on the risks of the bank and for credit exposures is likely to have further implications on corporate lending and the related cost of debt.[84]

## 12.9 General Conclusion of the Empirical Analysis

The empirical analysis covers the value relevance of stakeholder orientation in the first and public value topic salience in the second part. The first analysis shows the statistical significance of social stakeholders, such as society, community, and employees as well as issues covered by the SDG framework and linked to public value on sustainability and financial performance. It also shows that corporate sustainability disclosure appreciates a dedication towards the social side, showing a negative association of an orientation towards financial stakeholders such as banks.

[83] Hoepner et al. (2023); Mercereau et al. (2022); Neubaum and Zahra (2006).

[84] Galletta and Mazzù (2023); Houston and Shan (2022).

Social stakeholders, such as society, community, and employees, and the topic of quality education represented by SDG4 further reveal a negative association and a positive impact on the cost of capital.

Joining the discussion about organizational legitimacy, the empirical results verify the influence of societal discourse on organizational legitimacy and salient social topics from a public value perspective. Respecting existing limitations that, among others mentioned in this dissertation, relate to a limited maturity and standardization of corporate sustainability disclosure and the lack of clearly defined reporting standards, the illustrated results show that various public and private organizations such as standard setters, rating agencies, investors, banks, and companies have to face societal discourse. Furthermore, these players should create channels and dialogue to improve knowledge and transparency in describing the role of social stakeholders and their related issues in corporate reporting.

The results further present empirical evidence for contributions that have been made related to stakeholder theory and public value theory, revealing that social stakeholders matter for value considerations and that public value suits as a theory and framework for measuring organizational legitimacy from a societal perspective. In addition, it shows that accepted theories such as public value theory combined with more practical-oriented frameworks such as the SDGs can lead to significant results that generate new insights and knowledge about corporate reporting.

# Discussion and Outlook 13

The world, and as a crucial part of it, the business domain, is under the scrutiny of society and social actors that, in recent years, gained the power to become a judge of organizational legitimacy in deciding about a company's "social license to operate"[1]. Positively speaking, society and its related discourse have a sympathetic ear for the needs of the people, shaping the values of the public. For companies, the verification of the underlying hypotheses of this dissertation shows the need for acquiring a "contextual sensitivity"[2] that relates to public discourse and the discursive shifts that happen in society concerning the business world and its responsibilities.

The limitations of this dissertation acknowledge that non-financial reporting itself does not represent the mitigation factor for the deterioration and all adverse impacts that humans have on nature and related ecosystems. However, corporate disclosure is a reference for the strategic direction of companies, be it the orientation towards its stakeholders or the most relevant issues in focus. A more profound orientation toward social stakeholders and issues related to human needs is a prerequisite of corporate actions in today's world. Those issues being counteracted with the help of the SDGs incorporate a more detailed view of the organizational environment and its expectations, respecting the social stake in organizational legitimacy. As Pizzi (2021) put it, "the development of an accounting culture characterized by a real orientation toward sustainable development could represent a way to engage with stakeholders about the exigence to rethink the global economy"[3].

Referring to the primary goal of this dissertation, as mentioned in the introduction, and with the link between existing findings on organizational legitimacy and valuable research contributions relating to stakeholder theory and public value

[1] Boutilier and Thomson (2011), p. 4.

[2] Høvring et al. (2018), p. 641.

[3] Pizzi et al. (2021), p. 99., see Bebbington and Unerman (2018).



T. Heroth, *Corporate Sustainability Disclosure Matters*,
https://doi.org/10.1007/978-3-658-47653-3_13

theory, this dissertation can empirically verify the value relevance of stakeholder orientation and public value topic salience. It further sheds more light on the second objective by testing positive and negative associations of the orientation towards relevant social stakeholders with sustainability and financial performance and the sum of SGDs and single SDGs reported that provide evidence that a discursive shift towards *the social* in business took place. As a critical factor for establishing and maintaining organizational legitimacy, considering social stakeholders and their related issues must be part of the corporate strategic agenda, corporate communication, and sustainability disclosure. The empirical analysis further revealed that it is the moral-ethical dimension of public value that has an impact on sustainability and social performance. Consequently, focusing on issues such as inequalities present in life is worth counteracting, not only on a private but also on an organizational level. A goal that may not only be self-evident but also appreciated in the sustainability evaluation by rating agencies. In addition, a consideration and fostering of education reveals an effect on the cost of capital of a company. A domain that, especially in the developments towards a more sustainable society and business world, will further have a crucial position deemed to be fostered by organizations.

For the third goal of this dissertation that relates to its readers, some key aspects are further mentioned to trigger the motivation of regulatory bodies, standard setters, rating agencies, banks, investors, companies, and researchers.

With stricter ESG regulations, social stakeholders such as society, community, and employees will be the focus of sustainability disclosure. Consequently, corporate reports will be professionalized and further standardized. However, with greater attention toward social stakeholders and a continuously increasing urgency of social issues to be solved or at least counteracted, companies must focus on the pulse of topics that shape organizational legitimacy. Parallel to stricter regulations, standard setters are forced to provide structures and content on standards for sustainability reporting. The ESRS that the EU has published is the first development in this direction. The remaining work that must be accomplished relates to a deeper understanding of materiality and its practical implications. From a global perspective, a separation between promoters of the principle of single materiality and supporters of a double materiality approach can hardly be a final solution for corporate sustainability disclosure. Frameworks such as the SDGs promoted by the UN could further test, assess, and apply guiding structures that relate to the principle of double materiality. In line with this development, sustainability ratings will further align, consolidate, and become more transparent in what is needed for a company to improve or keep a targeted rating. Although these developments are challenging regarding resources, capacity, and expertise, they will require companies to keep pace with developments towards social sustainability, its actions, and the professionalization of sustainability disclosure.

Practitioners can embrace the empirical evidence that is being continuously added to the field of corporate sustainability reporting. With a more detailed view of the role of stakeholders, companies will be able to better report about their environment and the salient topics that evolve around their stakeholders. More maturity in corporate sustainability reporting is gained when the first reports directly include the content required by regulations, such as the CSRD, reported according to the ESRS. Therefore, researchers should dive deeper into constructs of interest in corporate reporting, providing further empirical evidence for well-accepted theories related to organizations, stakeholders, and the dynamics behind organizational legitimacy being shaped by societal discourse.

# References

Abdel-Maksoud, A., Jabbour, M., & Abdel-Kader, M. (2021). Stakeholder pressure, eco-control systems, and firms' performance: Empirical evidence from UK manufacturers *Accounting Forum*, *45*(1), 30–57. https://doi.org/10.1080/01559982.2020.1827697

Abson, D. J., Fischer, J., Leventon, J., Newig, J., Schomerus, T., Vilsmaier, U., von Wehrden, H., Abernethy, P., Ives, C. D., Jager, N. W., & Lang, D. J. (2017). Leverage points for sustainability transformation *Ambio*, *46*(1), 30–39. https://doi.org/10.1007/S13280-016-0800-Y/TABLES/2

Adams, C. A. (2004). The ethical, social and environmental reporting-performance portrayal gap *Accounting, Auditing & Accountability Journal*, *17*(5), 731–757. https://doi.org/10.1108/09513570410567791/FULL/PDF

Adams, C. A. (2017). *The Sustainable Development Goals, integrated thinking and the integrated report* (tech. rep.) Integrated Reporting. https://www.integratedreporting.org/wp-content/uploads/2017/09/SDGs-and-the-integrated-report%7B%5C_%7Dfull17.pdf

Adams, C. A., Alhamood, A., He, X., Tian, D. J., Wang, L., & Wang, Y. (2021). *The double-materiality concept—application and issues* (tech. rep.) GRI.

Adams, C. A., Druckman, P. B., & Picot, R. C. (2020). *Sustainable Development Goals Disclosure (SDGD) Recommendations* (tech. rep.) published by ACCA, Chartered Accountants ANZ, ICAS, IFAC, IIRC and WBA. https://www.ifac.org/publications/login/49449

Adler, P. S., & Kwon, S. W. (2002). Social capital: Prospects for a new concept Academy of Management. https://doi.org/10.5465/AMR.2002.5922314

Aiken, M., & Hage, J. (1968). Organizational Interdependence and Intra-Organizational Structure *American Sociological Review*, *33*(6), 912. https://doi.org/10.2307/2092683

Ajmal, M. M., Khan, M., Hussain, M., & Helo, P. (2018). Conceptualizing and incorporating social sustainability in the business world *International Journal of Sustainable Development & World Ecology*, *25*(4), 327–339. https://doi.org/10.1080/13504509.2017.1408714

Alexander, F. (2022). One Small Step From Financial Materiality to Sesquimateriality: A Critical Conceptual Leap for the ISSB The Harvard Law School Forum on Corporate Governance. https://doi.org/10.2139/SSRN.3908163

Alexander, S. (1920). *Space, Time and Deity: The Gifford Lectures at Glasgow 1916–1918 (Vol. 2)*. New York, The Humanities Press. https://archive.org/details/spacetimeanddeit00alexuoft?view=theater%7B%5C#%7Dpage/300/mode/2up%7B%5Cldots%7D



T. Heroth, *Corporate Sustainability Disclosure Matters*,
https://doi.org/10.1007/978-3-658-47653-3

Alford, J., & Hughes, O. (2008). Public Value Pragmatism as the Next Phase of Public Management *The American review of public administration*, *38*(2), 130–148. https://doi.org/10.1177/0275074008314203

Amiri, S., King, D. R., & Duesing, R. J. (2022). Managing divestments as projects: Benefits of stakeholder orientation *International Journal of Project Management*, *40*(4), 385–397. https://doi.org/10.1016/J.IJPROMAN.2022.03.012

Anandarajan, M., Hill, C., & Nolan, T. (2018). *Practical text analytics: Maximizing the value of text data* (2nd ed.) Springer International Publishing.

Arellano, M. (1993). On the testing of correlated effects with panel data *Journal of Econometrics*, *59*(1–2), 87–97. https://doi.org/10.1016/0304-4076(93)90040-C

Ashforth, B. E., & Gibbs, B. W. (1990). The Double-Edge of Organizational Legitimation *1*. (2), 177–194. https://doi.org/10.1287/ORSC.1.2.177

Audi, R., Loughran, T., & McDonald, B. (2016). Trust, but Verify: MD&A Language and the Role of Trust in Corporate Culture *Journal of Business Ethics*, *139*(3), 551–561. https://doi.org/10.1007/s10551-015-2659-4

Backhaus, K., Erichson, B., Plinke, W., & Weiber, R. (2018). Multivariate Analysemethoden *Multivariate Analysemethoden*. https://doi.org/10.1007/978-3-662-56655-8

Baier, P., Berninger, M., & Kiesel, F. (2020). Environmental, social and governance reporting in annual reports: A textual analysis *Financial Markets, Institutions & Instruments*, *29*(3), 93–118. https://doi.org/10.1111/fmii.12132

Banerjee, S. B. (2003). Who sustains whose development? Sustainable development and the reinvention of nature *Organization Studies*, *24*(1), 143–180. https://doi.org/10.1177/0170840603024001341

Bannier, C., Pauls, T., & Walter, A. (2019). Content analysis of business communication: Introducing a German dictionary *Journal of Business Economics*, *89*(1), 79–123. https://doi.org/10.1007/s11573-018-0914-8

Bañon Gomis, A. J., Guillén Parra, M., Hoffman, W. M., & Mcnulty, R. E. (2011). Rethinking the Concept of Sustainability *Business and Society Review*, *116*(2), 171–191. https://doi.org/10.1111/J.1467-8594.2011.00381.X

Barko, T., Cremers, M., & Renneboog, L. (2022). Shareholder Engagement on Environmental, Social, and Governance Performance *Journal of Business Ethics*, *180*(2), 777–812. https://doi.org/10.1007/S10551-021-04850-Z/TABLES/12

Barman, E. (2018). Doing well by doing good: A comparative analysis of ESG standards for responsible investment In *Advances in strategic management* (pp. 289–311). Emerald Group Publishing Ltd. https://doi.org/10.1108/S0742-332220180000038016

Barnett, W. P., & Carroll, G. R. (2003). Modeling Internal Organizational Change *Annual Review of Sociology*, *21*(1), 217–236.

Barrena Martínez, J., López Fernández, M., & Romero Fernández, P. M. (2016). Corporate social responsibility: Evolution through institutional and stakeholder perspectives *European Journal of Management and Business Economics*, *25*(1), 8–14. https://doi.org/10.1016/j.redee.2015.11.002

Bauer, R., Derwall, J., & Tissen, C. (2023). Private Shareholder Engagements on Material ESG Issues *SSRN Electronic Journal*. https://doi.org/10.2139/SSRN.4171496

Baumueller, J., & Sopp, K. (2022). Double materiality and the shift from non-financial to european sustainability reporting: Review, outlook and implications *Journal of Applied Accounting Research*, *23*(1), 8–28. https://doi.org/10.1108/JAAR-04-2021-0114

Bebbington, J., & Unerman, J. (2018). Achieving the United Nations Sustainable Development Goals: An enabling role for accounting research *Accounting, Auditing and Accountability Journal, 31*(1), 2–24. https://doi.org/10.1108/AAAJ-05-2017-2929/FULL/PDF

Bebchuk, L. A., Kastiel, K., & Tallarita, R. (2020). For Whom Corporate Leaders Bargain *SSRN Electronic Journal*. https://doi.org/10.2139/ssrn.3677155

Bebchuk, L. A., & Tallarita, R. (2020). The Illusory Promise of Stakeholder Governance *Cornell Law Review, 106*. https://heinonline.org/HOL/Page?handle=hein.journals/clqv106%7B%5C&%7Did=99%7B%5C&%7Ddiv=5%7B%5C&%7Dcollection=journals

Bechmann, G., & Stehr, N. (2002). The legacy of niklas luhmann *Society, 39*(2), 67–75. https://doi.org/10.1007/BF02717531/METRICS

Belsley, D. A. (1991). *Conditioning diagnostics: Collinearity and weak data in regression* New York, Wiley. https://cir.nii.ac.jp/crid/1130282272606256000

Benington, J., & Moore, M. (2010). *Public value: Theory and practice* Bloomsbury Publishing.

Benington, J., & Moore, M. H. (2011). *Public value in complex and changing times* London, Palgrave.

Benington, J. (2009). Creating the Public In Order To Create Public Value? *Intl Journal of Public Administration, 32*(3–4), 232–249. https://doi.org/10.1080/01900690902749578

Bennis, W. G. (1959). Leadership Theory and Administrative Behavior: The Problem of Authority *Administrative Science Quarterly, 4*(3), 259. https://doi.org/10.2307/2390911

Berg, F., Koelbel, J., & Rigobon, R. (2019). Aggregate Confusion: The Divergence of ESG Ratings *SSRN Electronic Journal*. https://doi.org/10.2139/ssrn.3438533

Berg, F., Kölbel, J., Pavlova, A., & Rigobon, R. (2021). ESG Confusion and Stock Returns: Tackling the Problem of Noise *SSRN Electronic Journal*. https://doi.org/10.2139/ssrn.3941514

Berger, P. L., & Luckmann, T. (1966). *The Social Construction of Reality: A Treatise in the Sociology of Knowledge* London, Penguin Books.

Berman, S. L., Wicks, A. C., Kotha, S., & Jones, T. M. (1999). Does Stakeholder Orientation Matter? The Relationship Between Stakeholder Management Models and Firm Financial Performance *Academy of Management Journal, 42*(5), 488–506. https://doi.org/10.5465/256972

Bernstein, L. A. (1967). The concept of materiality *The Accounting Review, 42*(1), 86–95.

Beske, F., Haustein, E., & Lorson, P. C. (2020). Materiality analysis in sustainability and integrated reports *Sustainability Accounting, Management and Policy Journal, 11*(1), 162–186. https://doi.org/10.1108/SAMPJ-12-2018-0343/FULL/PDF

Bettinazzi, E. L., & Feldman, E. R. (2021). Stakeholder orientation and divestiture activity *Academy of Management Journal, 64*(4), 1078–1096. https://doi.org/10.5465/AMJ.2019.0627

Bettinazzi, E. L., & Zollo, M. (2017). Stakeholder Orientation and Acquisition Performance *Strategic Management Journal, 38*(12), 2465–2485. https://doi.org/10.1002/SMJ.2672

Biermann, F., Hickmann, T., Sénit, C. A., Beisheim, M., Bernstein, S., Chasek, P., Grob, L., Kim, R. E., Kotzé, L. J., Nilsson, M., Ordóñez Llanos, A., Okereke, C., Pradhan, P., Raven, R., Sun, Y., Vijge, M. J., van Vuuren, D., & Wicke, B. (2022). Scientific evidence on the political impact of the Sustainable Development Goals *Nature Sustainability 2022 5:9, 5*(9), 795–800. https://doi.org/10.1038/s41893-022-00909-5

Biermann, F., Kanie, N., & Kim, R. E. (2017). Global governance by goal-setting: The novel approach of the UN Sustainable Development Goals *Current Opinion in Environmental Sustainability*, *26-27*, 26–31. https://doi.org/10.1016/J.COSUST.2017.01.010

Bilolo, C. (2018). *Legitimacy, Public Value, & Capital Allocation* (Doctoral Thesis) University of St. Gallen.

Bion, W. (1961). *Experience in Groups* New York, Routledge.

Bird, S., Klein, E., & Loper, E. (2009). *Natural Language Processing with Python* O'Reilly Media Inc. https://www.nltk.org/book/

Blair, M. M., & Stout, L. A. (1999). A Team Production Theory of Corporate Law. *Virginia Law Review*, *85*(2), 247–328.

Blau, P. M. (1957). Formal Organization: Dimensions of Analysis *American Journal of Sociology*, *63*(1), 58–69. https://doi.org/10.1086/222134

Blei, D. M., Ng, A. Y., & Edu, J. B. (2003). *Latent Dirichlet Allocation Michael I. Jordan* (tech. rep.) https://www.jmlr.org/papers/volume3/blei03a/blei03a.pdf?TB%7B%5C_%7Diframe=true%7B%5C&%7Dwidth=370.8%7B%5C&%7Dheight=658.8

Bloomberg L.P. (N.d.). Bloomberg ESG Scores.

Bonnafous-Boucher, M., & Rendtorff, J. D. (2016). *Stakeholder Theory* Cham, Springer International Publishing. https://doi.org/10.1007/978-3-319-44356-0

Boons, F., & Lüdeke-Freund, F. (2013). Business models for sustainable innovation: State-of-the-art and steps towards a research agenda *Journal of Cleaner Production*, *45*, 9–19. https://doi.org/10.1016/j.jclepro.2012.07.007

Boutilier, R. G., & Thomson, I. (2011). Modelling and measuring the social license to operate: Fruits of a dialogue between theory and practice retrieved. August 5, 2023, from https://www.socialicense.com/publications/Modelling%20and%20Measuring%20the%20SLO.pdf

Bowen, H. R. (N.d.). *Social responsibilities of the businessman* Iowa, University of Iowa Press.

Bozeman, B. (2007). Public values and public interest: Counterbalancing economic individualism *Georgetown University Press*.

Bramley, G., & Power, S. (2009). Urban Form and Social Sustainability: The Role of Density and Housing Type *Environment and Planning B: Urban Analytics and City Science*, *36*(1), 30–48. https://doi.org/10.1068/B33129

Brenner, S. N., & Molander, E. (1977). Is ethics of business changing *Harvard Business Review*, *55*(1), 57–71.

Breuer, W., Knetsch, A., & Salzmann, A. J. (2020). What Does It Mean When Managers Talk About Trust? *Journal of Business Ethics*, *166*(3), 473–488. https://doi.org/10.1007/s10551-018-4072-2

Briggs, L. (1968). Sequencing of Instruction in Relation to Hierarchies of Competence retrieved. August 5, 2023, from https://eric.ed.gov/?id=ED018975

Brown, J., & Jesse, D. (2015). Dialogic accountings for stakeholders: On opening up and closing down participatory governance *Journal of Management Studies*, *52*(7), 961–985. https://doi.org/10.1111/JOMS.12153

Buniamin, S. (2020). Exploring Managers' Perceptions on Stakeholders' Attributes in ESG Reporting of Malaysian Companies. *International Journal of Business Management (IJBM)*, *3*(1), 33–47.

Burchell, J., Cook, J., Burchell, J., & Cook, J. (2013). CSR, Co-optation and Resistance: The Emergence of New Agonistic Relations Between Business and Civil Society *Journal of Business Ethics, 115*(4), 741–754. https://doi.org/10.1007/S10551-013-1830-Z

Butterfield, K. D., Reed, R., & Lemak, D. J. (2004). An Inductive Model of Collaboration From the Stakeholder's Perspective *Business & Society, 43*(2), 162–195. https://doi.org/10.1177/0007650304265956

Capelle-Blancard, G. , & Petit, A. (2019). Every Little Helps? ESG News and Stock Market Reaction *Journal of Business Ethics, 157*(2), 543–565. https://doi.org/10.1007/s10551-017-3667-3

Carroll, A. B. (1979). A Three-Dimensional Conceptual Model of Corporate Performance *The Academy of Management Review, 4*(4), 497. https://doi.org/10.2307/257850

Carroll, A. B. (1991). The pyramid of corporate social responsibility: Toward the moral management of organizational stakeholders *Business Horizons, 34*(4), 39–48. https://doi.org/10.1016/0007-6813(91)90005-G

Carroll, A. B. (2008). A History of Corporate Social Responsibility: Concepts and Practices *The Oxford Handbook of Corporate Social Responsibility*. https://doi.org/10.1093/OXFORDHB/9780199211593.003.0002

Carroll, A. B., Brown, J. A., & Buchholtz, A. K. (2013). *Business & society: Ethics, sustainability, and stakeholder management* (10th ed.) Cengage Learning.

Carroll, G. R., & Hannan, M. T. (1989). Density Delay in the Evolution of Organizational Populations: A Model and Five Empirical Tests *34*. (3), 411. https://doi.org/10.2307/2393151

Chan, H. (2022). China moves to standardize fragmented ESG reporting landscape—Regulatory Intelligence Report retrieved. October 28, 2023, from https://www.thomsonreuters.com/en-us/posts/news-and-media/china-esg-reporting/

Chapin, F. S., Carpenter, S. R., Kofinas, G. P., Folke, C., Abel, N., Clark, W. C., Olsson, P., Smith, D. M. S., Walker, B., Young, O. R., Berkes, F., Biggs, R., Grove, J. M., Naylor, R. L., Pinkerton, E., Steffen, W., & Swanson, F. J. (2010). Ecosystem stewardship: Sustainability strategies for a rapidly changing planet *Trends in Ecology and Evolution, 25*(4), 241–249. https://doi.org/10.1016/j.tree.2009.10.008

Chiu, I. H. (2022). The eu sustainable finance agenda: Developing governance for double materiality in sustainability metrics *European Business Organization Law Review, 23*(1), 87–123. https://doi.org/10.1007/S40804-021-00229-9/METRICS

Christ, K. L., Burritt, R. L., Guthrie, J., & Evans, E. (2018). The potential for 'boundary-spanning organisations' in addressing the research-practice gap in sustainability accounting Emerald Group Holdings Ltd. https://doi.org/10.1108/SAMPJ-06-2017-0059

Churet, C., & Eccles, R. G. (2014). Integrated Reporting, Quality of Management, and Financial Performance *Journal of Applied Corporate Finance, 26*(1), 56–64. https://doi.org/10.1111/jacf.12054

Cicchiello, A. F., Marrazza, F., & Perdichizzi, S. (2023). Non-financial disclosure regulation and environmental, social, and governance (ESG) performance: The case of EU and US firms *Corporate Social Responsibility and Environmental Management, 30*(3), 1121–1128. https://doi.org/10.1002/CSR.2408

Coleman, J. S. (1990). Rational Organization, 2(1), 94–105. https://doi.org/10.1177/1043463190002001005

Consolandi, C., Eccles, R. G., & Gabbi, G. (2020). How material is a material issue? stock returns and the financial relevance and financial intensity of esg materiality *Journal of Sustainable Finance & Investment*, 1–24. https://doi.org/10.1080/20430795.2020.1824889

Cooper, D. J., & Morgan, W. (2013). Meeting the evolving corporate reporting needs of government and society: Arguments for a deliberative approach to accounting rule making *Accounting and Business Research*, *43*(4), 418–441. https://doi.org/10.1080/00014788.2013.794411

Cornell, B. , & Shapiro, A. C. (2021). Corporate stakeholders, corporate valuation and ESG *European Financial Management*, *27*(2), 196–207. https://doi.org/10.1111/eufm.12299

Costanza, R., Daly, L., Fioramonti, L., Giovannini, E., Kubiszewski, I., Mortensen, L. F., Pickett, K. E. , Ragnarsdottir, K. V., De Vogli, R., & Wilkinson, R. (2016). Modelling and measuring sustainable wellbeing in connection with the UN Sustainable Development Goals *Ecological Economics*, *130*, 350–355. https://doi.org/10.1016/J.ECOLECON.2016.07.009

Court, U. S. (1976). TSC Industries, Inc. v. Northway, Inc., 426 U.S. 438 retrieved. September 7, 2023, from https://supreme.justia.com/cases/federal/us/426/438/

Crilly, D., & Sloan, P. (2012). Enterprise logic: Explaining corporate attention to stakeholders from the 'inside-out' *Strategic Management Journal*, *33*(10), 1174–1193. https://doi.org/10.1002/SMJ.1964

Crilly, D., & Sloan, P. (2013). Autonomy or Control? Organizational Architecture and Corporate Attention to Stakeholders *Organization Science*, *25*(2), 339–355. https://doi.org/10.1287/ORSC.2013.0849

Cupertino, S., Vitale, G., & Ruggiero, P. (2022). Performance and (non) mandatory disclosure: The moderating role of the Directive 2014/95/EU *Journal of Applied Accounting Research*, *23*(1), 163–183. https://doi.org/10.1108/JAAR-04-2021-0115/FULL/PDF

Curtó-Pagès, F., Ortega-Rivera, E., Castellón-Durán, M., & Jané-Llopis, E. (2021). Coming in from the Cold: A Longitudinal Analysis of SDG Reporting Practices by Spanish Listed Companies Since the Approval of the 2030 Agenda *Sustainability 2021, Vol. 13, Page 1178*, *13*(3), 1178. https://doi.org/10.3390/SU13031178

Dando, N., & Swift, T. (2003). Transparency and Assurance: Minding the Credibility Gap *Journal of Business Ethics*, *44*(2–3), 195–200. https://doi.org/10.1023/A:1023351816790/METRICS

Deakin, S. (2005). The Coming Transformation of Shareholder Value *Corporate Governance: An International Review*, *13*(1), 11–18. https://doi.org/10.1111/J.1467-8683.2005.00399.X

Debenedetti, A., Philippe, D., Chaney, D., & Humphreys, A. (2021). Maintaining legitimacy in contested mature markets through discursive strategies: The case of corporate environmentalism in the French automotive industry *Elsevier*, *92*, 332–343.

Deetz, S., & Simpson, J. (2004). Critical organizational dialogue In *Dialogue: Theorizing difference in communication studies* (pp. 141–158). London, Sage.

Deloitte. (2022). *Corporate Sustainability Reporting Directive—The Future Landscape of Sustainability Reporting* (tech. rep.) Deloitte. https://www2.deloitte.com/content/dam/Deloitte/mt/Documents/sustainability/Corporate-Sustainability-Reporting-Directive-brochure-2022.pdf

Devlin, J., Chang, M. W., Lee, K., & Toutanova, K. (2019). BERT: Pre-training of deep bidirectional transformers for language understanding *NAACL HLT 2019–2019 Conference of the North American Chapter of the Association for Computational Linguistics: Human Language Technologies—Proceedings of the Conference*, *1*arXiv 1810.04805, 4171–4186.

Dhaliwal, D. S., Li, O. Z., Tsang, A., & Yang, Y. G. (2011). Voluntary nonfinancial disclosure and the cost of equity capital: The initiation of corporate social responsibility reporting *Accounting Review*, *86*(1), 59–100. https://doi.org/10.2308/accr.00000005

DiMaggio, P. J., & Powell, W. W. (1983). The Iron Cage Revisited: Institutional Isomorphism and Collective Rationality in Organizational Fields *American Sociological Review*, *48*(2), 147. https://doi.org/10.2307/2095101

Dobson, A. (1998). *Justice and the environment: Conceptions of environmental sustainability and theories of distributive justice* Oxford, Oxford University Press.

Donaldson, T., & Preston, L. E. (1995). The Stakeholder Theory of the Corporation: Concepts, Evidence, and Implications *Academy of Management Review*, *20*(1), 65–91. https://doi.org/10.5465/amr.1995.9503271992

Drucker, P. F. (1954). *The practice of management* New York, Harper & Row.

Drucker, P. F. (2001). *The Essential Drucker: The Best of Sixty Years of Peter Druckers Essential Writings on Management (Collins Business Essential Series)* HarperCollins. https://beech.wovelo.com/OPMw/oQZLZ/The%20Essential%20Drucker%20The%20Best%20of%20Sixty%20Years%20of%20Peter%20Druckers%20Essential%20Writings%20on%20Management%20Collins%20Business%20Essentials%20Free%20Book%20Online.pdf

Dryzek, J. S. (1999). Transnational democracy *The Journal of Political Philosophy*, *7*(1), 30–51.

Dryzek, J. S. (2001). Legitimacy and economy in deliberative democracy *Political Theory*, *29*(5), 651–669. https://doi.org/10.1177/0090591701029005003/ASSET/0090591701029005003.FP.PNG_V03

Durkheim, E. (1895). *The rules of sociological method* (S. Lukes, Ed.) New York, The Mcmillan Press Ltd.

Dworkin, R. (1978). Liberalism In S. Hampshire (Ed.), *Public and private moralit* New York, Cambridge University Pres.

Dyllick, T., & Hockerts, K. (2002). Beyond the business case for corporate sustainability *Business Strategy and the Environment*, *11*(2), 130–141. https://doi.org/10.1002/bse.323

Eberstadt, N. N. (1977). *What history tells us about corporate responsability* Boston, Little, Brown; Company. https://www.econbiz.de/Record/what-history-tells-us-about-corporate-responsability-eberstadt-nicholas/10002093979

Eccles, R. G., Krzus, M. P., Rogers, J., & Serafeim, G. (2012). The Need for Sector-Specific Materiality and Sustainability Reporting Standards *Journal of Applied Corporate Finance*, *24*(2), 65–71. https://doi.org/10.1111/J.1745-6622.2012.00380.X

Eccles, R. G., & Serafeim, G. (2013). The performance frontier: Innovating for a sustainable strategy retrieved. September 8, 2023, from https://hbr.org/2013/05/the-performance-frontier-innovating-for-a-sustainable-strategy

Edgley, C. (2014). A genealogy of accounting materiality *Critical Perspectives on Accounting*, *25*(3), 255–271. https://doi.org/10.1016/J.CPA.2013.06.001

EFRAG. (2022a). *Draft European Sustainability Reporting Standards—Explanatory Note* (tech. rep.) EFRAG. https://www.efrag.org/sites/default/files/sites/webpublishing/SiteAssets/Appendix%202.6%20-%20WP%20on%20draft%20ESRG%201.pdf

EFRAG. (2022b). *Draft European Sustainability Reporting Standards—Explanatory Note* (tech. rep.) EFRAG. https://www.efrag.org/sites/default/files/sites/webpublishing/SiteAssets/03%20Explanatory%20note%20Fist%20set%20of%20ESRS%20Article%2029%20b_last.pdf

EFRAG. (2023). *06–02 Materiality Assessment SRB 230823* (tech. rep.) EFRAG. https://www.efrag.org/system/files/sites/webpublishing/Meeting%20Documents/2307280747599961/06-02%20Materiality%20Assessment%20SRB%20230823.pdf

El Ghoul, S., Guedhami, O., Kwok, C. C., & Mishra, D. R. (2011). Does corporate social responsibility affect the cost of capital? *Journal of Banking and Finance*, *35*(9), 2388–2406. https://doi.org/10.1016/j.jbankfin.2011.02.007

Eliwa, Y., Aboud, A., & Saleh, A. (2021). ESG practices and the cost of debt: Evidence from EU countries *Critical Perspectives on Accounting*, *79*, 102097. https://doi.org/10.1016/j.cpa.2019.102097

Elkington, J. (1997). The triple bottom line In M. V. Russo (Ed.), *Environmental management: Readings and cases* (pp. 49–66).

Epstein, A. (1989). *Mind, fantasy, and healing* New York, Delacorte.

Epstein, S. (2003). Cognitive-experiential self theory of personality In T. Millon, M. L. Lerner, & I. B. Weiner (Eds.), *Handbook of psychology* (5th ed., pp. 159–184). New York, John Wiley & Sons, Inc.

EURLex. (2019). Guidelines on non-financial reporting: Supplement on reporting climate-related information retrieved. September 7, 2023, from https://eur-lex.europa.eu/legal-content/EN/TXT/PDF/?uri=CELEX:52019XC0620(01)

European Commission. (2023). A European Green Deal—Striving to be the first climate-neutral continent retrieved. September 5, 2023, from https://commission.europa.eu/strategy-and-policy/priorities-2019-2024/european-green-deal%7B%5C_%7Den

European Parliament. (2023). Corporate Sustainability Reporting Directive (CSRD)—In "A European Green Deal" retrieved. September 5, 2023, from https://www.europarl.europa.eu/legislative-train/theme-a-european-green-deal/file-review-of-the-non-financial-reporting-directive

European Union. (2014). Directive 2014/95/EU of the European Parliament and of the Council of 22 October 2014 amending Directive 2013/34/EU as regards disclosure of non-financial and diversity information by certain large undertakings and groups retrieved. September 5, 2023, from https://eur-lex.europa.eu/legal-content/EN/TXT/?uri=celex%7B%5C%25%7D3A32014L0095

European Union. (2023). Official journal of the european union: Corporate social responsibility directive retrieved. September 4, 2023, from https://eur-lex.europa.eu/legal-content/EN/TXT/?uri=CELEX:32022L2464

Fatemi, A., Glaum, M., & Kaiser, S. (2018). ESG performance and firm value: The moderating role of disclosure *Global Finance Journal*, *38*, 45–64. https://www.sciencedirect.com/science/article/pii/S1044028316300680

Fløttum, K., & Gjerstad, Ø. (2017). Narratives in climate change discourse *Wiley Interdisciplinary Reviews: Climate Change*, *8*(1), e429. https://doi.org/10.1002/WCC.429

Fonseca, L., Domingues, J., & Dima, A. (2020). Mapping the sustainable development goals relationships *mdpi.com*, *12*(3359), 1–15. https://doi.org/10.3390/su12083359

Frederick, W. C. (1960). The growing concern over business responsibility *California Management Review*, *2*(4), 54–61. http://dx.doi.org/10.2307/41165405

Freeman, R. E. (1984). *Strategic management: A stakeholder approach* Boston, Pitman.
Freeman, R. E. (1994). The Politics of Stakeholder Theory: Some Future Directions *Business Ethics Quarterly*, *4*(4), 409–421. https://doi.org/10.2307/3857340
Freeman, R. E. (1999). Divergent stakeholder theory *Academy of Management Review*, *24*(2), 233–236. https://doi.org/10.5465/AMR.1999.1893932
Freeman, R. E. (2010). Managing for stakeholders: Trade-offs or value creation *Journal of Business Ethics*, *96*(1), 7–9. https://doi.org/10.1007/s10551-011-0935-5
Freeman, R. E., Phillips, R., & Sisodia, R. (2020). Tensions in Stakeholder Theory *Business & Society*, *59*(2), 213–231. https://doi.org/10.1177/0007650318773750
Freudenreich, B., Lüdeke-Freund, F., & Schaltegger, S. (2020). A Stakeholder Theory Perspective on Business Models: Value Creation for Sustainability *Journal of Business Ethics*, *166*(1), 3–18. https://doi.org/10.1007/s10551-019-04112-z
Friede, G., Busch, T., & Bassen, A. (2015). ESG and financial performance: Aggregated evidence from more than 2000 empirical studies *Journal of Sustainable Finance and Investment*, *5*(4), 210–233. https://doi.org/10.1080/20430795.2015.1118917
Gabriel, R., & Sailer, U. (2023). Development of the concept of sustainability In *Sustainable business management* (pp. 31–35). Tübingen, UVK Verlag.
Galletta, S., & Mazzù, S. (2023). ESG controversies and bank risk taking *Business Strategy and the Environment*, *32*(1), 274–288. https://doi.org/10.1002/BSE.3129
Gao, G. (2023). Law Over Borders Comparative Guide: Environmental, Social & Governance retrieved. September 6, 2023, from https://www.globallegalpost.com/lawoverborders/environmental-social-governance-559211028/china-1543409311
George, G., Howard-Grenville, J., Joshi, A., & Tihanyi, L. (2016). Understanding and Tackling Societal Grand Challenges through Management Research *Academy of Management Journal*, *59*(6), 1880–1895. https://doi.org/10.5465/AMJ.2016.4007
Giese, G., Lee, L. E., Melas, D., Nagy, Z., & Nishikawa, L. (2019). Foundations of ESG investing: How ESG affects equity valuation, risk, and performance *Journal of Portfolio Management*, *45*(5), 69–83. https://doi.org/10.3905/jpm.2019.45.5.069
Glozer, S., Hibbert, S., & Caruana, R. (2013). *From Monologue to Dialogue: Mapping Dialogical Traditions within Co-Creation and Corporate Social Responsibility (CSR)* (tech. rep.) Nottingham University Business School. http://www.nottingham.ac.uk/business/ICCSR
Goloshchapova, I., Poon, S. H., Pritchard, M., & Reed, P. (2019). Corporate social responsibility reports: Topic analysis and big data approach *European Journal of Finance*, *25*(17), 1637–1654. https://doi.org/10.1080/1351847X.2019.1572637
Gomez, P. (1981). Modelle und Methoden des systemorientierten Managements *Bern*, Haupt Verlag.
Gomez, P., & Probst. (1999). *Die Praxis des ganzheitlichen Probleml ö sens Vernetzt denken—Unternehmerisch handeln—Pers ö nlich ü berzeugen.* Bern, Haupt Verlag.
Gorddard, R., Colloff, M. J., Wise, R. M., Ware, D., & Dunlop, M. (2016). Values, rules and knowledge: Adaptation as change in the decision context *Environmental Science & Policy*, *57*, 60–69. https://doi.org/10.1016/J.ENVSCI.2015.12.004
Governance and Accountability Institute. (2020). *Flash Report S&P 500 2020* (tech. rep.) Governance and Accountability Institute, Inc. https://www.ga-institute.com/research-reports/flash-reports/2020-sp-500-flash-report.html

GRI. (2015). Item 21—Transition to GRI Standards retrieved. September 11, 2023, from https://www.globalreporting.org/standards/media/2336/item-21-transition-to-gri-standards-mock-up-of-sustainability-reporting-standard-3-quality-principles.pdf

GRI. (2020a). GRI Universal Standards: GRI 101, GRI 102, and GRI 103—Exposure draft retrieved. April 6, 2020, from https://www.globalreporting.org/standards/media/2605/universal-exposure-draft.pdf 2020-04-06

GRI. (2020b). *Linking the SDGs and the GRI Standards* (tech. rep.) GRI. https://www.globalreporting.org/media/lbvnxb15/mapping-sdgs-gri-update-march.pdf

GRI. (2023). GRI 2: General Disclosures 2021 retrieved. April 1, 2023, from https://www.globalreporting.org/pdf.ashx?id=12358

GRI. (N.d.). GSSB FAQ retrieved. September 11, 2023, from https://www.globalreporting.org/standards/global-sustainability-standards-board/gssb-faq/

GRI, & EFRAG. (2023). Joint Press Release from EFRAG and the Global Reporting Iniative (GRI) retrieved. September 9, 2023, from https://www.globalreporting.org/media/gjsmbi2b/press-release-gri-efrag-joint-statement-v-230901.pdf

GRI, & Support the Goals. (2022). *State of Progress: Business Contributions to the SDGs* (tech. rep.) Global Reporting Initiative, Support the Goals. https://www.globalreporting.org/media/ab5lun0h/stg-gri-report-final.pdf

Guignon, C., & Schnadelbach, H. (1984). Philosophy in Germany, 1831–1933 *The Philosophical Review*, *96*(1), 112. https://doi.org/10.2307/2185333

Guo, L., Vargo, C. J., Pan, Z., Ding, W., & Ishwar, P. (2016). Big social data analytics in journalism and mass communication: Comparing dictionary-based text analysis and unsupervised topic modeling *Journalism and Mass Communication Quarterly*, *93*(2), 322–359. https://doi.org/10.1177/1077699016639231

Haber, S. (1964). *Efficiency and Uplift: Scientific Management in the Progressive Era 1890-1920* Chicago, University of Chicago Press.

Habermas, J. (1996). *Between Facts and Norms: Contributions to a Discourse Theory of Law and Democracy* Cambridge, MA, The MIT Press.

Habermas, J. (1998). Between Facts and Norms: An Author's Reflections *Denver University Law Review*, *76*.

Habermas, J. (1987). *The Theory of Communicative Action—Lifeworld and System: A Critique of Functionalist Reason* (2nd ed.) Boston, Beacon Press.

Hammond, M. (2021). Sustainability as a cultural transformation: The role of deliberative democracy *The Political Prospects of a Sustainability Transformation*, 173–192. https://doi.org/10.4324/9781003132288-10

Hannan, M. T., & Freeman, J. (1989). Organizational Ecology *Organizational Ecology*. https://doi.org/10.2307/J.CTVJZ813K

Hansch, D. (1997). *Psychosynergetik: Die fraktale Evolution des Psychischen. Grundlagen einer Allgemeinen Psychotherapie* Opladen, Germany, Westdeutscher Verlag.

Harper Ho, V. (2020). Disclosure overload? lessons for risk disclosure & esg reporting reform from the regulation s-k concept release *Villanova Law Review*, *65*(1). https://digitalcommons.law.villanova.edu/vlr/vol65/iss1/2

Harrison, J. S., Bosse, D. A., & Phillips, R. A. (2010). Managing for stakeholders, stakeholder utility functions, and competitive advantage *Strategic Management Journal*, *31*(1), 58–74. https://doi.org/10.1002/SMJ.801

Hartley, J., Alford, J., Knies, E., & Douglas, S. (2017). Towards an empirical research agenda for public value theory *Public Management Review*, *19*(5), 670–685. https://doi.org/10.1080/14719037.2016.1192166

Heath, R. L., Pearce, W. B., Shotter, J., Taylor, J. R., Kersten, A., Zorn, T., Roper, J., Motion, J., & Deetz, S. (2006). The processes of dialogue: Participation and legitimation *Management Communication Quarterly*, *19*(3), 341–375. https://doi.org/10.1177/0893318905282208

Henry, E., & Leone, J. A. (2016). Measuring qualitative information in capital markets research: Comparison of alternative methodologies to measure disclosure tone American Accounting Association. https://doi.org/10.2308/accr-51161

Hermes, F. (2021). *Finanzielle und kommunikative Manipulation im M&A-Kontext: Empirische Evidenz zu internationalen Zielunternehmen* (Doctoral Thesis) Universität St. Gallen.

Herrfahrdt-Pähle, E., Schlüter, M., Olsson, P., Folke, C., Gelcich, S., & Pahl-Wostl, C. (2020). Sustainability transformations: Socio-political shocks as opportunities for governance transitions *Global Environmental Change*, *63*, 102097. https://doi.org/10.1016/J.GLOENVCHA.2020.102097

Heyde, J. E. (1926). *Eine philosophische Grundlegung* Erfurt, Germany, Verlag Kurt Stenger.

Hildebrand, P., Bodnar, P., Crozar, C., & Velikova, E. (2021). *Sustainable Investing. Integrating the UN SDGs in Investments.* (tech. rep.) Blackrock. https://www.blackrock.com/institutions/en-zz/literature/whitepaper/integrating-un-sdgs-investments.pdf

Hirschman, A. O. (1982). *Hirschman, A. O.* Princeton, Princeton University Press.

Hoepner, A. G. F., Oikonomou, I., Sautner, Z., Starks, L. T., & Zhou, X. (2023). *ESG Shareholder Engagement and Downside Risk* (tech. rep.) Swiss Finance Institute Research Paper. Elsevier BV. https://doi.org/10.2139/SSRN.2874252

Hoffman, A. J., & Ross, S. M. (1999). Working Paper Institutional Evolution and Change: Environmentalism and the US Chemical Industry *Academy of Management Journal*, *42*(4), 351–371. http://ssrn.com/abstract=2940277

Hörisch, J., Freeman, R. E., & Schaltegger, S. (2014). Applying Stakeholder Theory in Sustainability Management *Organization & Environment*, *27*(4), 328–346. https://doi.org/10.1177/1086026614535786

Hörisch, J., Schaltegger, S., & Freeman, R. E. (2020). Integrating stakeholder theory and sustainability accounting: A conceptual synthesis *Journal of Cleaner Production*, *275*, 124097. https://doi.org/10.1016/j.jclepro.2020.124097

Houston, J. F., & Shan, H. (2022). Corporate ESG Profiles and Banking Relationships *The Review of Financial Studies*, *35*(7), 3373–3417. https://doi.org/10.1093/RFS/HHAB125

Høvring, C. M., Andersen, S. E., & Nielsen, A. E. (2018). Discursive Tensions in CSR Multistakeholder Dialogue: A Foucauldian Perspective *Journal of Business Ethics*, *152*(3), 627–645. https://doi.org/10.1007/S10551-016-3330-4/FIGURES/4

Hutton, W., Clements, J., & Sang, B. (2007). Public Leadership for the 21st Century Delivering 'Public Value' through Entrepreneurship, Engagement and Rigour *International Journal of Leadership in Public Services*, *3*(1), 47–51. https://doi.org/10.1108/17479886200700006

ICLG. (N.d.). Environmental, Social & Governance Law 2023 retrieved. September 6, 2023, from https://iclg.com/practice-areas/environmental-social-and-governance-law/usa

Ike, M., Donovan, J. D., Topple, C., & Masli, E. K. (2019). The process of selecting and prioritising corporate sustainability issues: Insights for achieving the Sustainable Development

Goals *Journal of Cleaner Production*, *236*, 117661. https://doi.org/10.1016/J.JCLEPRO.2019.117661

Imerman, D. (2018). Contested Legitimacy and Institutional Change: Unpacking the Dynamics of Institutional Legitimacy *International Studies Review*, *20*(1), 74–100. https://doi.org/10.1093/ISR/VIX039

International Financial Regulation Standards. (2022). Ifrs foundation completes consolidation with value reporting foundation retrieved. September 8, 2023, from https://www.ifrs.org/news-and-events/news/2022/08/ifrs-foundation-completes-consolidation-with-value-reporting-foundation/

Isaacs, W. N. (1993). Taking flight: Dialogue, collective thinking, and organizational learning *Organizational dynamics*, *22*(2), 24–39. https://www.sciencedirect.com/science/article/pii/0090261693900512

Iwin, A. A. (1975). *Grundlagen der Logik von Wertungen* Berlin, Akademie-Verlag.

Izzo, M. F., Ciaburri, M., & Tiscini, R. (2020). The Challenge of Sustainable Development Goal Reporting: The First Evidence from Italian Listed Companies *Sustainability 2020, Vol. 12, Page 3494*, *12*(8), 3494. https://doi.org/10.3390/SU12083494

Jalbert, J. E. (1988). Husserl's Position Between Dilthey and the Windelband-Rickert School of Neo-Kantianism *Journal of the History of Philosophy*, *26*(2), 279–296. https://doi.org/10.1353/HPH.1988.0045

Jaworska, S., & Nanda, A. (2016). Doing Well by Talking Good: A Topic Modelling-Assisted Discourse Study of Corporate Social Responsibility *Applied Linguistics*, *39*(3), amw014. https://doi.org/10.1093/applin/amw014

Jebe, R. (2019). The convergence of financial and esg materiality: Taking sustainability mainstream *American Business Law Journal*, *56*(3), 645–702. https://doi.org/10.1111/ablj.12148

Jeffery, N. (2009). *Stakeholder engagement: A road map to meaningful engagement* (tech. rep.) Doughty Centre, Cranfield School of Management. https://www.fundacionseres.org/lists/informes/attachments/1118/stakeholder%20engagement.pdf

Johnson, H. L. (1971). *Business in contemporary society: framework and issues* Belmont, California, Wadsworth Pub. Co.

Jones, T. M. (1980). Corporate Social Responsibility Revisited, Redefined *California Management Review*, *22*(3), 59–67. https://doi.org/10.2307/41164877

Jones, T. M. (1995). Instrumental Stakeholder Theory: A Synthesis of Ethics and Economics in The Academy of Management Review *The Academy of Management Review*, *20*(2), 404–437. https://doi.org/258852

Jones, T. M. (2011). The nature of firm-stakeholder relationships: Realizing the potential of an underappreciated contribution of Freeman's 25-year-old classic In R. A. Phillips (Ed.), *Stakeholder theory* (pp. 54–76). Cheltenham, Edward Elgar Publishing. https://books.google.ch/books?hl=en%7B%5C&%7Dlr=%7B%5C&%7Did=6TjaV8vNZO4C%7B%5C&%7Doi=fnd%7B%5C&%7Dpg=PA54%7B%5C&%7Ddq=Jones,+T.+M.+2011.+The+nature+of+firm-stakeholder+relationships:+Realizing+the+potential+of+an+underappreciated+contribution+of+Freeman' s+25-year-old+classic.+In+R.+Phillips+(Ed.),+

Jones, T. M., Harrison, J. S., & Felps, W. (2018). How applying instrumental stakeholder theory can provide sustainable competitive advantage *Academy of Management Review*, *43*(3), 371–391. https://doi.org/10.5465/AMR.2016.0111

Jonker, J., & Nijhof, A. (2006). Looking through the eyes of others: Assessing mutual expectations and experiences in order to shape dialogue and collaboration between business and NGOs with respect to CSR *Corporate Governance: An International Review*, *14*(5), 456–466. https://doi.org/10.1111/J.1467-8683.2006.00518.X

Kant, I. (1785). *Grundlegung zur Metaphysik der Sitten* (AA, Ed.; 4th ed.).

Katz, D., & Kahn, R. (1966). *The Social Psychology of Organizations* Hoboken, John Wiley; Sons Inc.

Kezdi, G. (2004). Robust standard error estimation in fixed-effects panel models *papers.ssrn.com*, *88*(4). https://papers.ssrn.com/sol3/papers.cfm?abstract%7B%5C_%7Did=596988

Kim, B., Kim, E., & Foss, N. J. (2016). Balancing absorptive capacity and inbound open innovation for sustained innovative performance: An attention-based view *European Management Journal*, *34*(1), 80–90. https://doi.org/10.1016/j.emj.2015.10.002

Kim, Y., Park, M. S., & Wier, B. (2012). Is Earnings Quality Associated with Corporate Social Responsibility? *The Accounting Review*, *87*(3), 761–796. https://doi.org/10.2308/ACCR-10209

Kirkby, J., O'Keefe, P., & Timberlake, L. (1995). *The Earthscan Reader in Sustainable Development* Taylor & Francis.

Kloppenburg, J. T. (1986). *Uncertain Victory: Social Democracy and Progressivism in European and American Thought 1870-1920* Oxford, Oxford University Press.

Knebel, S., & Seele, P. (2015). Quo vadis gri? a (critical) assessment of gri 3.1 a+ non-financial reports and implications for credibility and standardization *Corporate Communications*, *20*(2), 196–212. https://doi.org/10.1108/CCIJ-11-2013-0101/FULL/PDF

Korca, B., & Costa, E. (2020). Directive 2014/95/EU: Building a research agenda *Journal of Applied Accounting Research*, *22*(3), 401–422. https://doi.org/10.1108/JAAR-05-2020-0085/FULL/PDF

Kotler, P., & Lee, N. (2008). *Corporate social responsibility: Doing the most good for your company and your cause* John Wiley & Sons.

Kotsantonis, S., & Serafeim, G. (2019). Four Things No One Will Tell You About ESG Data *Journal of Applied Corporate Finance*, *31*(2), 50–58. https://doi.org/10.1111/JACF.12346

Koumbarakis, A., Offenhammer, A., Jaccard, S., Blumer, R., Dermaku, M., & Balli, J. (2023). *CSRD: Revolution instead of evolution* (tech. rep.) PWC. Zurich. https://www.pwc.ch/en/insights/fs/csrd-revolution-instead-of-evolution.html

KPMG. (2022). *Big shifts, small steps -Survey of Sustainability Reporting* (tech. rep.) KPMG. https://assets.kpmg/content/dam/kpmg/xx/pdf/2023/04/big-shifts-small-steps.pdf

KPMG. (2023). Corporate Sustainability Reporting Directive—Is your organization ready for the Corporate Sustainability Reporting Directive? Retrieved. September 5, 2023, from https://kpmg.com/nl/en/home/topics/environmental-social-governance/corporate-sustainability-reporting-directive.html

Krippendorff, K. (2011). Agreement and information in the reliability of coding *Communication Methods and Measures*, *5*(2), 93–112. https://doi.org/10.1080/19312458.2011.568376

Kuhlman, T., & Farrington, J. (2010). What is Sustainability? *Sustainability 2010, Vol. 2, Pages 3436–3448*, *2*(11), 3436–3448. https://doi.org/10.3390/SU2113436

Kuratko, D. F., Hornsby, J. S., & Goldsby, M. G. (2007). The Relationship of Stakeholder Salience, Organizational Posture, and Entrepreneurial Intensity to Corporate Entrepreneur-

ship *Journal of Leadership & Organizational Studies*, *13*(4), 56–72. https://doi.org/10.1177/107179190701300408 01
Kwartler, T. (2017). *Text Mining in Practice with R* Chichester, UK, John Wiley & Sons, Ltd. https://doi.org/10.1002/9781119282105
La Torre, M., Sabelfeld, S., Blomkvist, M., & Dumay, J. (2020). Rebuilding trust: Sustainability and non-financial reporting and the European Union regulation *Meditari Accountancy Research*, *28*(5), 701–725. https://doi.org/10.1108/MEDAR-06-2020-0914/FULL/XML
Landi, G., & Sciarelli, M. (2019). Towards a more ethical market: The impact of ESG rating on corporate financial performance *Social Responsibility Journal*, *15*(1), 11–27. https://doi.org/10.1108/SRJ-11-2017-0254
Lankoski, L. (2016). Alternative conceptions of sustainability in a business context *Journal of Cleaner Production*, *139*, 847–857. https://doi.org/10.1016/J.JCLEPRO.2016.08.087
Leal Filho, W. (2000). Dealing with misconceptions on the concept of sustainability *International Journal of Sustainability in Higher Education*, *1*(1), 9–19. https://doi.org/10.1108/14676300010307066/FULL/HTML
Lee, J. C. Y. (2021). China Publishes Environmental and Social Disclosure Rules for Listed Companies retrieved. September 6, 2023, from https://www.mayerbrown.com/en/perspectives-events/blogs/2021/07/china-publishes-environmental-and-social-disclosure-rules-for-listed-companies
Leng, B., Hua, J., & Lloyd, S. (2023). ISSB to include GRI and ESRS in IFRS S1 sources of guidance retrieved. September 9, 2023, from https://www.corporatedisclosures.org/content/top-stories/issb-to-include-gri-and-esrs-in-ifrs-s1-sources-of-guidance.html
Linklaters. (2022). Climate-related disclosure may soon become mandatory in the United States retrieved. September 6, 2023, from https://www.linklaters.com/en/knowledge/publications/alerts-newsletters-and-guides/2022/march/23/climate-related-disclosure-may-soon-become-mandatory-in-the-united-states
Littig, B., & Griessler, E. (2005). Social sustainability: A catchword between political pragmatism and social theory *International Journal of Sustainable Development*, *8*(1–2), 65–79. https://doi.org/10.1504/IJSD.2005.007375
Lokuwaduge, C. S. S., & De Silva, K. M. (2022). ESG Risk Disclosure and the Risk of Green Washing *Australasian Accounting, Business and Finance Journal*, *16*(1), 146–159. https://doi.org/10.14453/aabfj.v16i1.10
Long, J. S., & Ervin, L. H. (2000). Using Heteroscedasticity Consistent Standard Errors in the Linear Regression Model Statistical Computing and Graphics Using Heteroscedasticity Consistent Standard Errors in the Linear Regression Model *54*. (3), 217–224.
Lounsbury, M., & Glynn, M. A. (2001). Cultural entrepreneurship: Stories, legitimacy, and the acquisition of resources *Strategic Management Journal*, *22*(6–7), 545–564. https://doi.org/10.1002/SMJ.188
Lovins, L. H. (2016). Needed: A Better Story *Humanistic Management Journal*, *1*(1), 75–90. https://doi.org/10.1007/S41463-016-0004-4/METRICS
Lüdeke-Freund, F., & Dembek, K. (2017). Sustainable business model research and practice: Emerging field or passing fancy? *Journal of Cleaner Production*, *168*, 1668–1678. https://doi.org/10.1016/j.jclepro.2017.08.093
Luhmann, N. (2012). *Theory of society, volume 1* Stanford University Press. https://books.google.ch/books?hl=en%7B%5C&%7Dlr=%7B%5C&%7Did=xt%7B%5C_%7DLPvc7fX8C%7B%5C&%7Doi=fnd%7B%5C&%7Dpg=PT9%7B%5C&

%7Ddq=luhmann+theory+of+society%7B%5C&%7Dots=BqyYHVt5oo%7B%5C&%7Dsig=HOpOiYSGXNEey-oxBv5MCQXopMg

Maak, T., & Pless, N. M. (2006). *Responsible leadership* London, England, Routledge.

March, J. G. (1996). Continuity and Change in Theories of Organizational Action *Administrative Science Quarterly, 41*(2), 278–287.

Marschlich, S., & Ingenhoff, D. (2021). Stakeholder engagement in a multicultural context: The contribution of (personal) relationship cultivation to social capital *Public Relations Review, 47*(4), 102091. https://www.sciencedirect.com/science/article/pii/S0363811121000849

Marshall, J. D., & Toffel, M. W. (2005). Framing the elusive concept of sustainability: A sustainability hierarchy *Environmental Science and Technology, 39*(3), 673–682. https://doi.org/10.1021/ES040394K/ASSET/IMAGES/LARGE/ES040394KF00003.JPEG

Martz, W. (2013). Evaluating Organizational Performance: Rational, Natural, and Open System Models *American Journal of Evaluation, 34*(3), 385–401. https://doi.org/10.1177/1098214013479151

Marx, K. (1865). Value, Price and Profit *Marx Today*, 99–122. https://doi.org/10.1057/9780230117457_7

Mayo, E. (1949). *Hawthorne and the western electric company. The social problems of an industrial civilisation* https://www.practicesurvival.com/wa_files/Hawthorne_20Studies_201924_20Elton_20Mayo.pdf

Mayring, P. (2019). Qualitative content analysis: Demarcation, varieties, developments *In Forum: Qualitative Social Research, Freie Universit ä t Berlin, 20*(3). https://www.qualitative-research.net/index.php/fqs/article/download/3343/4557?inline=1

Mayring, P. (2014). *Qualitative Content Analysis Theoretical Foundation, Basic Procedures and Software Solution* (tech. rep.) AUT. www.beltz.de

Mead, G. H. (1934). *Mind, self & society* (C. W. Morris, Ed.) Chicago, University of Chicago Press.

Menger, C. (1871). *Grunds ä tze der Volkswirtschaftslehre: Erster, Allgemeiner Teil.* Wien, Wilhelm Braumüller K.K. Hof und Universitätsbuchhändler.

Mercereau, B., Melin, L., Lugo, M. M., Mercereau, B., Melin, L., & Lugo, M. M. (2022). Creating shareholder value through ESG engagement *Journal of Asset Management, 23*(7), 550–566. https://doi.org/10.1057/S41260-022-00270-4

Merriam-Webster. (N.d.-a). Existing definitions of society retrieved. August 15, 2023, from https://www.merriam-webster.com/dictionary/society

Merriam-Webster. (N.d.-b). Existing definitions of value retrieved. August 23, 2023, from https://www.merriam-webster.com/dictionary/value

Merton, R. K. (1968). *Social Theory and Social Structure* New York, The Free Press.

Messier, W. F., Martinov-Bennie, N., & Eilifsen, A. (2005). A Review and Integration of Empirical Research on Materiality: Two Decades Later *AUDITING: A Journal of Practice & Theory, 24*(2), 153–187. https://doi.org/10.2308/AUD.2005.24.2.153

Meyer, J. W., & Rowan, B. (1977). Institutionalized Organizations: Formal Structure as Myth and Ceremony *American Journal of Sociology, 83*(2), 340–363. https://about.jstor.org/terms

Meynhardt, T., Gomez, P., & Berndt, T. (2017). How non-financial reporting can make use of Public Value: Getting Serious About Society *The Reporting Times, 25*.

Meynhardt, T. (2009). Public Value Inside: What is Public Value Creation? *International Journal of Public Administration, 32*(3-4), 192–219. https://doi.org/10.1080/01900690902732632

Meynhardt, T. (2015). Public Value: Turning a Conceptual Framework into a Scorecard In J. M. BRYSON, B. C. CROSBY, & L. BLOOMBERG (Eds.), *Public management and change series. public value and public administration* (pp. 147–169). Washington, DC, Georgetown University Press.

Meynhardt, T. (2019). Public value: Value creation in the eyes of society In *Public value* (pp. 5–22). Routledge. https://doi.org/10.4324/9781315163437-1

Meynhardt, T., & Bäro, A. (2019). Public value reporting: Adding value to (non-)financial reporting *Public Value*, 87–108. https://doi.org/10.4324/9781315163437-7

Meynhardt, T., Brieger, S. A., & Hermann, C. (2020). Organizational public value and employee life satisfaction: The mediating roles of work engagement and organizational citizenship behavior *The International Journal of Human Resource Management, 31*(12), 1560–1593. https://doi.org/10.1080/09585192.2017.1416653

Meynhardt, T., & Gomez, P. (2019). Building Blocks for Alternative Four-Dimensional Pyramids of Corporate Social Responsibilities *Business and Society, 58*(2), 404–438. https://doi.org/10.1177/0007650316650444/ASSET/IMAGES/LARGE/10.1177_0007650316650444-FIG4.JPEG

Miles, S. (2017). Stakeholder theory classification: A theoretical and empirical evaluation of definitions *Journal of Business Ethics, 142*(3), 437–459. https://doi.org/10.1007/s10551-015-2741-y

Milne, M. J., & Adler, R. W. (1999). Exploring the reliability of social and environmental disclosures content analysis *Accounting, Auditing & Accountability Journal, 12*(2), 237–256. https://doi.org/10.1108/09513579910270138

Mitchell, R. K., Agle, B. R., & Wood, D. J. (1997). Toward a Theory of Stakeholder Identification and Salience: Defining the Principle of Who and What Really Counts *The Academy of Management Review, 22*(4), 853. https://doi.org/10.2307/259247

Montiel, I., & Delgado-Ceballos, J. (2014). Defining and Measuring Corporate Sustainability: Are We There Yet? *Organization and Environment, 27*(2), 113–139. https://doi.org/10.1177/1086026614526413/ASSET/IMAGES/LARGE/10.1177_1086026614526413-FIG2.JPEG

Moore, M. H. (1995). Creating public value: Strategic management in government *Harvard university press.*

Moore, M. L., Olsson, P., Nilsson, W., Rose, L., & Westley, F. R. (2018). Navigating emergence and system reflexivity as key transformative capacities *Ecology and Society, 23*(2), 38–43. https://doi.org/10.5751/ES-10166-230238

Moore, M. H. (2013). *Recognizing Public Value* Harvard University Press. https://doi.org/10.4159/harvard.9780674067820

Morelli, J. (2011). Environmental Sustainability: A Definition for Environmental Professionals *Journal of Environmental Sustainability, 1*(1), 2. https://doi.org/10.14448/jes.01.0002

Moreno, A., & Redondo, T. (2016). Text Analytics: The convergence of Big Data and Artificial Intelligence *IJIMAI, 3*(6), 57–64. https://doi.org/10.9781/ijimai.2016.369

Morrell, K. (2009). GOVERNANCE AND THE PUBLIC GOOD *Public Administration, 87*(3), 538–556. https://doi.org/10.1111/J.1467-9299.2009.01756.X

Mulgan, G., Breckon, J., Tarrega, M., Bakhshi, H., Davies, J., Khan, H., & Finnis, A. (2019). *Public value: How can it be measured, managed and grown?* (Tech. rep.) Nesta. www.nesta.org.uk

Neubaum, D. O., & Zahra, S. A. (2006). Institutional Ownership and Corporate Social Performance: The Moderating Effects of Investment Horizon, Activism, and Coordination *Journal of Management*, *32*(1), 108–131. https://doi.org/10.1177/0149206305277797

Ng, A. C., & Rezaee, Z. (2015). Business sustainability performance and cost of equity capital *Journal of Corporate Finance*, *34*, 128–149. https://doi.org/10.1016/J.JCORPFIN.2015.08.003

Nietzsche, F. (1887). *On the Genealogy of Morality* Indiana, Hackett Publishing Company.

Nilsen, A. G. (2016). Power, Resistance and Development in the Global South: Notes Towards a Critical Research Agenda *International Journal of Politics, Culture and Society*, *29*(3), 269–287. https://doi.org/10.1007/S10767-016-9224-8

Nilsson, M., Chisholm, E., Griggs, D., Howden-Chapman, P., McCollum, D., Messerli, P., Neumann, B., Stevance, A. S., Visbeck, M., & Stafford-Smith, M. (2018). Mapping interactions between the sustainable development goals: Lessons learned and ways forward *Sustainability Science*, *13*(6), 1489–1503. https://doi.org/10.1007/S11625-018-0604-Z

Nolan, J. (2005). The United Nations' Compact with business: Hindering or helping the protection of human rights *University of Queensland Law Journal*, *24*(2), 445–466.

Owen, D. L. (1990). Towards a theory of social investment: A review essay *Accounting, Organizations and Society*, *15*(3), 249–265. https://doi.org/10.1016/0361-3682(90)90007-H

Palazzo, G., & Scherer, A. G. (2006). Corporate legitimacy as deliberation: A communicative framework *Journal of Business Ethics*, *66*(1), 71–88. https://doi.org/10.1007/S10551-006-9044-2/METRICS

Palmer, T. B., & Flanagan, D. J. (2016). The sustainable company: Looking at goals for people, planet and profits *Journal of Business Strategy*, *37*(6), 28–38. https://doi.org/10.1108/JBS-09-2015-0095/FULL/PDF

Parmar, B. L., Freeman, R. E., Harrison, J. S., Wicks, A. C., Purnell, L., & de Colle, S. (2010). Stakeholder Theory: The State of the Art *Academy of Management Annals*, *4*(1), 403–445. https://doi.org/10.5465/19416520.2010.495581

Parsons, T. (1956). Suggestions for a Sociological Approach to the Theory of Organizations.II *Administrative Science Quarterly*, *1*(2), 225. https://doi.org/10.2307/2390988

Perello-Marin, M. R., Rodriguez-Rodriguez, R., & Alfaro-Saiz, J. J. (2022). Analysing GRI reports for the disclosure of SDG contribution in European car manufacturers *Technological Forecasting and Social Change*, *181*, 121744. https://doi.org/10.1016/J.TECHFORE.2022.121744

Perrini, F., & Tencati, A. (2006). Sustainability and stakeholder management: The need for new corporate performance evaluation and reporting systems *Business Strategy and the Environment*, *15*(5), 296–308. https://doi.org/10.1002/BSE.538

Pettigrew, A. M. (1987). Context and Action in the Transformation of the Firm *Journal of Management Studies*, *24*(6), 649–670.

Pfeffer, J., & Salancik, G. R. (1978). *The External Control of Organizations: A Resource Dependence Perspective* Routledge. https://papers.ssrn.com/abstract=1496213

Philippe, D., & Durand, R. (2011). The impact of norm-conforming behaviors on firm reputation *Strategic Management Journal*, *32*(9), 969–993. https://doi.org/10.1002/SMJ.919

Phillips, N., Lawrence, T. B., & Hardy, C. (2004). Discourse and Institutions *Academy of Management Review*, *29*(4), 636–652.

Pizzi, S., Mara, I., Baldo, D., Caputo, F., & Venturelli, A. (2021). Voluntary disclosure of Sustainable Development Goals in mandatory non-financial reports: The moderating role of cultural dimension *Wiley Online LibraryS Pizzi, M Del Baldo, F Caputo, A VenturelliJournal of International Financial Management & Accounting, 2022•Wiley Online Library*, *00*(1), 1–24. https://doi.org/10.1111/jifm.12139

Posner, B. Z., & Schmidt, W. H. (1984). Values and the American Manager: An Update *California Management Review*, *26*(3), 202–216. https://doi.org/10.2307/41165090

Pulino, S. C., Ciaburri, M., Magnanelli, B. S., & Nasta, L. (2022). Does ESG Disclosure Influence Firm Performance? *Sustainability 2022, Vol. 14, Page 7595*, *14*(13), 7595. https://doi.org/10.3390/SU14137595

Puroila, J., & Mäkelä, H. (2019). Matter of opinion: Exploring the socio-political nature of materiality disclosures in sustainability reporting *Accounting, Auditing and Accountability Journal*, *32*(4), 1043–1072. https://doi.org/10.1108/AAAJ-11-2016-2788/FULL/PDF

Raimo, N., Caragnano, A., Zito, M., Vitolla, F., & Mariani, M. (2021). Extending the benefits of ESG disclosure: The effect on the cost of debt financing *Corporate Social Responsibility and Environmental Management*, *28*(4), 1412–1421. https://doi.org/10.1002/csr.2134

Rappaport, A. (1981). Selecting strategies that create shareholder value *Harvard Business Review*, *59*(3).

Rasouli, A. H. , & Kumarasuriyar, D. A. (2016). The Social Dimention of Sustainability: Towards Some Definitions and Analysis *Journal of Social Science for Policy Implications*, *4*(2). https://doi.org/10.15640/JSSPI.V4N2A3

Rees, W. E. (2021). Achieving Sustainability: Reform or Transformation? *The Earthscan Reader in Sustainable Cities*, 22–52. https://doi.org/10.4324/9781315800462-3

Refinitiv. (2022). *Environmental, Social and Governance Scores from Refinitiv* (tech. rep.) Refinitiv. https://www.refinitiv.com/content/dam/marketing/en%7B%5C_%7Dus/documents/methodology/refinitiv-esg-scores-methodology.pdf

Reidel, J. (2010). *Erfolgreich oder ruin? transnationale unternehmen und nachhaltige entwicklung kritische reflexion aus menschenrechtlicher perspektive. mchen* oekom.

Rescher, N. (1969). *Introduction to Value Theory* Englewood Cliffs, N.J., Prentice- Hall.

Rescher, N. (1982). *Introduction to value theory* Washington, DC, University Press of America.

Ricardo, D., & McCulloch, J. R. (1888). *The works of David Ricardo: With a notice of the life and writings of the Author* Murray.

Rodrigue, M., Magnan, M., & Boulianne, E. (2013). Stakeholders' influence on environmental strategy and performance indicators: A managerial perspective *Management Accounting Research*, *24*(4), 301–316. https://doi.org/10.1016/j.mar.2013.06.004

Rokeach, M. (1968). *Beliefs, Attitudes, and Values: A Theory of Organization and Change* San Francisco, CA, Jossey-Bass Publishers.

Roloff, J. (2008). Learning from multi-stakeholder networks: Issue-focussed stakeholder management *Journal of Business Ethics*, *82*(1), 233–250. https://doi.org/10.1007/s10551-007-9573-3

Rosati, F., & Faria, L. G. D. (2019). Business contribution to the Sustainable Development Agenda: Organizational factors related to early adoption of SDG reporting *Corporate Social*

*Responsibility and Environmental Management*, *26*(3), 588–597. https://doi.org/10.1002/CSR.1705

Rose, J., Flak, L. S., & Sæbø, Ø. (2018). Stakeholder theory for the E-government context: Framing a value-oriented normative core *Government Information Quarterly*, *35*(3), 362–374. https://doi.org/10.1016/j.giq.2018.06.005

Rüegg-Stürm, J., & Grand, S. (2019). *Managing in a complex world—The St. Gallen management model* Uni-Taschenbuecher GmbH (UTB).

Ryan, G., & Bernard, H. (2000). Data management and analysis methods In N. K. Denzin & Y. S. Lincoln (Eds.). Thousand Oaks, Sage.

Saad, A. I., & Strauss, D. (2020). A new "reasonable investor" and changing frontiers of materiality: Increasing investor reliance on esg disclosures and implications for securities litigation *Berkeley Business Law Journal*, *17*. https://heinonline.org/HOL/Page?handle=hein.journals/berkbusj17%7B%5C&%7Did=398%7B%5C&%7Ddiv=14%7B%5C&%7Dcollection=journals

Saffer, A. J. (2019). Fostering social capital in an international multi-stakeholder issue network *Public Relations Review*, *45*(2), 282–296. https://doi.org/10.1016/j.pubrev.2019.02.004

Sagoff, M. (1986). Values and Preferences *Ethics*, *96*(2), 301–316. https://doi.org/10.1086/292748

Sajwan, R., & Priya, C. (2018). How to test normality in STATA? https://www.projectguru.in/normality-test-stata/

Sami, A., Jusoh, A., Qureshi, M. I., & Skudai, M. (2016). Does Ethical Leadership Create Public Value? Empirical Evidences from Banking Sector of Pakistan *International Review of Management and Marketing*, *6*(4), 262–270. https://dergipark.org.tr/en/pub/irmm/issue/32091/355304

Schaltegger, S., Christ, K. L., Wenzig, J., & Burritt, R. L. (2022). Corporate sustainability management accounting and multi-level links for sustainability—A systematic review *International Journal of Management Reviews*. https://doi.org/10.1111/ijmr.12288

Schaltegger, S., Hörisch, J., & Freeman, R. E. (2019). Business Cases for Sustainability: A Stakeholder Theory Perspective *Organization & Environment*, *32*(3), 191–212. https://doi.org/10.1177/1086026617722882

Scheler, M. (1926). *Die Wissensformen und die Gesellschaft* Leipzig, Neue Geist Verl.

Schot, J., & Steinmueller, W. E. (2018). Three frames for innovation policy: R&D, systems of innovation and transformative change *Research Policy*, *47*(9), 1554–1567. https://doi.org/10.1016/J.RESPOL.2018.08.011

Schuetz, A. (1962). *The Problem of Social Reality: Collected Papers I*. The Hague, Martinus Nijhoff.

Schulzke, K. S., & Berger-Walliser, G. (2017). Toward a unified theory of materiality in securities law *Columbia Journal of Transnational Law*, *56*(6). https://heinonline.org/hol-cgi-bin/get%7B%5C_%7Dpdf.cgi?handle=hein.journals/cjtl56%7B%5C&%7Dsection=5%7B%5C&%7Dcasa%7B%5C_%7Dtoken=bzaL7xUpOXsAAAAA:sU8fqfEu%7B%5C_%7De6y8vIhAW%7B%5C_%7DD6JsxtKNJ9sZFDDk6fsacbu42%7B%5C_%7DCN%7B%5C_%7DDTbl9iDAVHPJUx-P1FF-5XvaXDI

Scott, W. R., & Davis, G. F. (2007). *Organizations and Organizing: Rational, Natural, and Open Systems Perspectives*. Upper Saddle River, New Jersey, Pearson Prentice Hall.

Scott, W. (1987). *Organizations: Rational, Natural, and Open Systems* (2nd ed.) Englewood Cliffs, N.J., Prentice Hall.

Scott, W. (2003). *Organizations: Rational, natural, and open systems* Upper Saddle River, NJ, Pearson.
Selznick, P. (1948). Foundations of the Theory of Organization *American Sociological Review*, *13*(1), 25. https://doi.org/10.2307/2086752
She, C., & Michelon, G. (2019). Managing stakeholder perceptions: Organized hypocrisy in CSR disclosures on Facebook *Critical Perspectives on Accounting*, *61*, 54–76. https://doi.org/10.1016/j.cpa.2018.09.004
Shenhav, Y. (1995). From Chaos to Systems: The Engineering Foundations of Organization Theory, 1879–1932 *Administrative Science Quarterly*, *40*(4), 557. https://doi.org/10.2307/2393754
Smith, R. W. (1970). The Concept of Legitimacy *Theoria: A Journal of Social and Political Theory*, (35), 17–29.
Sopp, K., & Baumueller, J. (2019). Nichtfinanzielle berichterstattung: Kritik an den neuen leitlinien zu klimabezogenen angaben *Der Betrieb*, *72*(33), 1801–1809.
Steinmann, H., & Scherer, A. G. (1998). Corporate Ethics and Global Business: Philosophical Considerations on Intercultural Management In B. N. Kumar & H. Steinmann (Eds.), *Ethics in international business* (pp. 13–46). Berlin/New York, Walter de Gruyter. https://papers.ssrn.com/abstract=1553111
Stern, R. N., & Barley, S. R. (1996). Organizations and social systems: Organization theory's neglected mandate *Administrative Science Quarterly*, *41*(1), 146–162. https://doi.org/10.2307/2393989
Stock, J. H., & Watson, M. W. (2008). Heteroskedasticity-robust standard errors for fixed effects panel data regression *Econometrica*, *76*(1), 155–174. https://doi.org/10.1111/J.0012-9682.2008.00821.X
Stocker, F., Arruda, M. P., Mascena, K. M. C., & Boaventura, J. M. G. (2020). Stakeholder engagement in sustainability reporting: A classification model *Corporate Social Responsibility and Environmental Management*, *27*(5), 2071–2080. https://doi.org/10.1002/csr.1947
Stockholm Resilience Center, S. U. (2016). The SDGs wedding cake https://www.stockholmresilience.org/research/research-news/2016-06-14-the-sdgs-wedding-cake.html
Strand, R. (1983). A systems paradigm of organizational adaptations to the social environment *Academy of management review. Academy of Management*, *8*(1), 90–96. https://doi.org/10.5465/AMR.1983.4287703
Suchman, M. C. (1995). Managing Legitimacy: Strategic and Institutional Approaches *The Academy of Management Review*, *20*(3), 571. https://doi.org/10.2307/258788
SustainAbility. (2020). *Rate the Raters 2020: Investor Survey and Interview Results* (tech. rep.) ERM Group Company. https://www.sustainability.com/globalassets/sustainability.com/thinking/pdfs/sustainability-ratetheraters2020-report.pdf
Swiss Sustainable Finance. (2023). Switzerland, sustainability regulations retrieved. September 6, 2023, from https://www.sustainablefinance.ch/en/resources/regulation/switzerland.html
Székely, F., & Knirsch, M. (2005). Responsible Leadership and Corporate Social Responsibility: Metrics for Sustainable Performance *European Management Journal*, *23*(6), 628–647. https://doi.org/10.1016/J.EMJ.2005.10.009

Szennay, Á., Szigeti, C., Kovács, N., & Szabó, D. R. (2019). Through the Blurry Looking Glass—SDGs in the GRI Reports *Resources 2019, Vol. 8, Page 101*, *8*(2), 101. https://doi.org/10.3390/RESOURCES8020101

Talbot, C. (2009). Public Value—The Next "Big Thing" in Public Management? *International Journal of Public Administration*, *32*(3-4), 167–170. https://doi.org/10.1080/01900690902772059

Talbot, C. (2011). Paradoxes and prospects of 'public value' *Public Money and Management*, *31*(1), 27–34. https://doi.org/10.1080/09540962.2011.545544

Taylor, F. (1911). *The Principles of Scientific Management* New York, Harper; Brothers.

Theodoulidis, B., Diaz, D., Crotto, F., & Rancati, E. (2017). Exploring corporate social responsibility and financial performance through stakeholder theory in the tourism industries *Tourism Management*, *62*, 173–188. https://doi.org/10.1016/j.tourman.2017.03.018

Thompson, J. D. (1967). *Organizations in Action* New York, McGraw-Hill.

Titscher, S., & Jenner, B. (2000). *Methods of text and discourse analysis: In search of meaning* Sage.

Tönnies, F., & Loomis, C. P. (1887). *Community and Society* Taylor & Francis.

Ulrich, H. (1968). *Die Unternehmung als produktives soziales System* (2., überar) Bern, Haupt Verlag.

Ulrich, H., & Probst, G. B. (1988). *Anleitung zum ganzheitlichen Denken und Handeln* Bern, Haupt Verlag.

United Nations. (1999). Kofi Annan's address to World Economic Forum in Davos retrieved. September 8, 2023, from https://www.un.org/sg/en/content/sg/speeches/1999-02-01/kofi-annans-address-world-economic-forum-davos

United Nations. (2023). United Nations Global Compact Participants retrieved. July 21, 2023, from https://unglobalcompact.org/what-is-gc/participants

United Nations. (N.d.). Universal Declaration of Human Rights retrieved. August 5, 2023, from https://www.un.org/en/about-us/universal-declaration-of-human-rights

United Nations Environment Programme Finance Initiative. (2004). *Who Cares Wins—Connecting Financial Markets to a Changing World* (tech. rep.) United Nations. https://www.unepfi.org/fileadmin/events/2004/stocks/who%7B%5C_%7Dcares%7B%5C_%7Dwins%7B%5C_%7Dglobal%7B%5C_%7Dcompact%7B%5C_%7D2004.pdf

United Nations Principles for Responsible Investment. (2022). PRI Regulation Database retrieved. September 5, 2023, from https://www.unpri.org/policy/regulation-database

United Nations Principles for Responsible Investment. (2023). Anti-ESG bills found unified opposition in US statehouses retrieved. September 6, 2023, from https://www.unpri.org/pri-blog/anti-esg-bills-found-unified-opposition-in-us-statehouses/11658.article

Value Reporting Foundation (SASB Standards). (2022). Materiality Finder retrieved. April 14, 2022, from https://www.sasb.org/standards/materiality-finder/find/

Velte, P. (2017). Does ESG performance have an impact on financial performance? Evidence from Germany *Journal of Global Responsibility*, *8*(2), 169–178. https://doi.org/10.1108/jgr-11-2016-0029

Verney, P. (2023). EU banking body says 'appropriate' that ISSB moves towards double materiality retrieved. September 8, 2023, from https://www.responsible-investor.com/eu-banking-body-says-appropriate-that-issb-moves-towards-double-materiality/

Vitolla, F., Raimo, N., Rubino, M., & Garzoni, A. (2019a). How pressure from stakeholders affects integrated reporting quality *Corporate Social Responsibility and Environmental Management*, *26*(6), csr.1850. https://doi.org/10.1002/csr.1850

Vitolla, F., Raimo, N., Rubino, M., & Garzoni, A. (2019b). The impact of national culture on integrated reporting quality. A stakeholder theory approach *Business Strategy and the Environment*, *28*(8), 1558–1571. https://doi.org/10.1002/bse.2332

von Carlowitz, H. C. (1713). *Silvicultura Oeconomica* (J. F. Braun, Ed.) Leipzig.

Waddock, S. (2020). Achieving sustainability requires systemic business transformation *Global Sustainability*, *3*(e12), 1–12. https://doi.org/10.1017/SUS.2020.9

Walsh, J. P. (2005). Book Review Essay: Taking Stock of Stakeholder Management *Academy of Management Review*, *30*(2), 426–438. https://doi.org/10.5465/AMR.2005.16387898

Weber, H. (2017). Politics of 'Leaving No One Behind': Contesting the 2030 Sustainable Development Goals Agenda *Globalizations*, *14*(3), 337–342.

Weber, M. (1922). Wirtschaft und Gesellschaft— Grundriß der verstehenden Soziologie Tuebinge, Germany. http://www.textlog.de/weber%7B%5C_%7Dwirtschaft.html

Weber, M. (1978). Basic Sociological Terms In G. Roth & C. Wittich (Eds.), *Economy and society. an outline of interpretive sociology* Berkeley, California. http://anthropos-lab.net/wp/wp-content/uploads/2011/12/Weber-%7B%5C%%7D0ABasic-Sociological-Terms.pdf

Wei, J., Tay, Y., Bommasani, R., Raffel, C., Zoph, B., Borgeaud, S., Yogatama, D., Bosma, M., Zhou, D., Metzler, D., Chi, E. H., Hashimoto, T., Vinyals, O., Liang, P., Dean, J., & Fedus, W. (2022). Emergent Abilities of Large Language Models arXiv 2206.07682v2. http://arxiv.org/abs/2206.07682

Weick, K. E., & Quinn, R. E. (2003). Oraganizational Change and Development *Annual Review of Psychology*, *50*, 361–386.

Weingaertner, C., & Moberg, Å. (2014). Exploring Social Sustainability: Learning from Perspectives on Urban Development and Companies and Products *Sustainable Development*, *22*(2), 122–133. https://doi.org/10.1002/SD.536

Wiersum, K. F. (1995). 200 years of sustainability in forestry: Lessons from history *Environmental Management*, *19*(3), 321–329. https://doi.org/10.1007/BF02471975/METRICS

Wood, S., Jones, S., Johnson, J., Brauman, K. A., Chaplin-Kramer, R., Fremier, A., Girvetz, E., Gordon, L. J., Kappel, C. V., Mandle, L., Mulligan, M., O'Farrell, P., Smith, W. K., Willemen, L., Zhang, W., & DeClerck, F. A. (2018). Distilling the role of ecosystem services in the Sustainable Development Goals *Ecosystem Services*, *70-82*.

Wördenweber, M. (2017). *Nachhaltigkeitsmanagement: Grundlagen und Praxis unternehmerischen Handelns* (10249th ed.) Schäffer-Poeschel Verlag.

World Business Council for Sustainable Development. (2018). *Corporate reporting in the United States and Canada* (tech. rep.) World Business Council for Sustainable Development. https://www.wbcsd.org/Programs/Redefining-Value/Resources/Corporate-reporting-in-the-United-States-and-Canada

World Commission and Development. (1983). *UN Document Report: The Brundtland Report "Our Common Future"* (tech. rep.) United Nations.

Yang, M., Evans, S., Vladimirova, D., & Rana, P. (2017). Value uncaptured perspective for sustainable business model innovation *Journal of Cleaner Production*, *140*, 1794–1804. https://doi.org/10.1016/j.jclepro.2016.07.102

Yuan, X., Li, Z., Xu, J., & Shang, L. (2022). ESG disclosure and corporate financial irregularities—Evidence from Chinese listed firms *Journal of Cleaner Production*, *332*, 129992. https://doi.org/10.1016/J.JCLEPRO.2021.129992

Yuchtman, E., & Seashore, S. E. (1967). A System Resource Approach to Organizational Effectiveness *American Sociological Review*, *32*(6), 891. https://doi.org/10.2307/2092843

Zollo, M., Minoja, M., Casanova, L., Hockerts, K., Neergaard, P., Schneider, S., & Tencati, A. (2009). Towards an internal change management perspective of CSR: Evidence from project RESPONSE on the sources of cognitive alignment between managers and their stakeholders, and their implications for social performance *Corporate Governance: The international journal of business in society*, *9*(4), 355–372. https://doi.org/10.1108/14720700910984918

Zumente, I., & Bistrova, J. (2021). ESG Importance for Long-Term Shareholder Value Creation: Literature vs. Practice *Journal of Open Innovation: Technology, Market, and Complexity 2021, Vol. 7, Page 127*, *7*(2), 127. https://doi.org/10.3390/JOITMC7020127

The manufacturer's authorised representative in the EU is Springer
Nature Customer Service Centre GmbH, Europaplatz 3, 69115 Heidelberg,
Germany. If you have any concerns regarding our products, please
contact ProductSafety@springernature.com

Printed and bound by CPI Group (UK) Ltd, Croydon, CR0 4YY
30/09/2026
02225566-0001